INTERNATIONAL SERIES OF MONOGRAPHS IN

NATURAL PHILOSOPHY

GENERAL EDITOR: D. TER HAAR

VOLUME 15

LIQUID METALS

OTHER TITLES IN THE SERIES
IN NATURAL PHILOSOPHY

LIQUID METALS

BY

N. H. MARCH

Laboratory of Atomic and Solid State Physics,
Cornell University, Ithaca,
New York, U.S.A.

THE QUEEN'S AWARD
TO INDUSTRY 1966

PERGAMON PRESS

OXFORD · LONDON · EDINBURGH · NEW YORK
TORONTO · SYDNEY · PARIS · BRAUNSCHWEIG

111586

546.3
M 315

Pergamon Press Ltd., Headington Hill Hall, Oxford
4 & 5 Fitzroy Square, London W.1
Pergamon Press (Scotland) Ltd., 2 & 3 Teviot Place, Edinburgh 1
Pergamon Press Inc. 44–01 21st Street, Long Island City, New York 11101
Pergamon of Canada, Ltd., 6 Adelaide Street East, Toronto, Ontario
Pergamon Press (Aust.) Pty. Ltd., Rushcutters Bay, Sydney, N.S.W.
Pergamon Press S.A.R.L., 24 rue des Écoles, Paris 5ᵉ
Vieweg & Sohn GmbH, Burgplatz 1, Braunschweig

First edition 1968

This book was written during the session 1965–6 while the author was on leave from the Department of Physics, The University, Sheffield.
During this period, the author wishes to acknowledge partial contractual support from the U.S. Atomic Energy Commission.

Library of Congress Catalog Card No. 67–27720

08 003229 X

Contents

Contents

Preface

MANY people have helped, directly or indirectly, in clarifying my ideas on the method of presentation finally adopted in this volume. Especial thanks are due to my friends and colleagues in the Physics Department at Sheffield University, and in particular to Drs. J. E. Enderby, T. Gaskell, W. Jones, G. E. Kilby, W. H. Young and Mr. M. J. Stott. I also wish to acknowledge the hospitality I received during my stay in the Laboratory of Atomic and Solid State Physics at Cornell University, and the stimulus I derived from this visit. I am particularly grateful to Drs. N. W. Ashcroft and R. C. Desai, as well as to Dr. P. Schofield of A.E.R.E., Harwell, for valuable comments. My wife helped me greatly with the final preparation of the book, and by typing the entire manuscript.

In a book of this size, it has not proved possible to give complete derivations of all the results used. I have tried, nevertheless, until the last two chapters, to write the book at such a level that only elementary quantum mechanics and statistical thermodynamics are required to understand it. When I have had to make a choice, I have preferred to give derivations centring round electron theory, as the classical statistical mechanics of fluids has been discussed so extensively elsewhere. In spite of this bias, I have not been able to do justice to pseudopotential theory in this book. Fortunately, this field has been covered fully in the recent book by Professor W. A. Harrison (1966), to which the reader is referred.

CHAPTER 1

Outline

THE purpose of this small volume is to present a unified, if sometimes oversimplified, picture of the properties of liquid metals. The emphasis is placed on a microscopic description of the electron states and the consequences which stem from such a point of view. However, experimental results are used both to suggest major features which the theory must incorporate, as well as to test the theory at every stage.

While much of the theory is undoubtedly at a primitive stage in its development, it seemed worth while to attempt to bring together an approach based on electron theory and the vast volume of work on classical statistical mechanics of fluids. It may also be, though it is premature to try to reach any final conclusion on this point, that liquid metals will eventually offer us a medium for extracting fully quantitative interatomic forces which will then be of considerable value in discussing solids. For example, a theory of the crystal structures of metals could be one of the rewards for exploring further the forces in liquid metals (Worster and March, 1964).

Though the book has primarily to do with conducting fluids, we have not hesitated to discuss, sometimes at length, the properties of insulating fluids, when it turns out to be valuable to compare and contrast the properties of one with the other.

The outline of the book is as follows. In the next chapter we discuss the quantitative way in which the structure of a monatomic fluid may be described, via the radial distribution function. This is conveniently tackled by considering the scattering of X-

1

rays (or neutrons) from the fluid. Chapter 3 concerns itself with the way in which the forces operating in liquid metals may be qualitatively described. This necessitates some introduction to the theory of electron screening in metals, and an account is given of the shielding of a structureless ion core in the Born approximation. Ways in which the structure of the ion core can profoundly influence the shielding are then described.

We next turn to consider whether the description afforded by electron theory is compatible with the structural characteristics of the metals as summarized in Chapter 2. This is done by essentially enquiring whether the experimental results can be put in a form where they reflect rather directly the interionic forces. It is shown that an important tool for doing this is the so-called direct correlation function introduced by Ornstein and Zernike long ago. Chapter 5 deals with what is known so far about the forces operating between ions in liquid metals. This brings us next to enquire whether the nature of the forces we have been discussing can lead us to a theory of melting. This obviously involves some brief discussion of the properties of solid metals, particularly the role of phonons and vacancies, and these matters are dealt with in Chapter 6.

The topic of electron scattering and resistivity of a liquid metal is then, quite naturally, taken up and the simple but remarkably successful theory of Ziman and his collaborators is fully described. Chapter 8 deals with the time dependent generalization of the structure factor and a discussion is given of the information we can obtain on the dynamics of fluids from inelastic neutron scattering. The link with conductivity is discussed in Appendix 6.

Finally, in Chapter 9 we consider in a little more detail the nature of the energy level spectrum of the electrons in a liquid metal. The general concepts to be used in such a discussion have been laid down, particularly by Edwards (1962), but the problem of obtaining realistic results in liquid metals is truly formidable and little general agreement exists among workers in the field. Nevertheless, some preliminary attempts to calculate dispersion relations and densities of states have been made, and these are summarized.

CHAPTER 2

Liquid structure

IT WAS Debye (1915) who first demonstrated that in the discussion of X-ray diffraction in liquids, one had to consider two atoms whose scattered rays interfere with one another, and as a result an interference pattern is obtained which depends crucially on the relative separation of the two atoms, or the pair function $g(r)$ introduced below (see especially eqn. (2.13)).

While, in principle, $g(r)$ could be calculated from statistical mechanics (see Chapters 4 and 5), and hence the X-ray intensity patterns predicted, the major progress in understanding the structure of liquids has come from analysis of the X-ray (or neutron) data, to yield $g(r)$. The approach adopted below to relate the X-ray intensity to $g(r)$, follows more closely the work of Warren and Gingrich (1934; see especially Gingrich 1943) than the original Debye argument.

2.1 Debye's formula

We consider an X-ray beam, of amplitude E_0, propagating along the direction of the unit vector s_0 shown in Fig. 1. Suppose the polarization to be such that the electric field is normal to the plane of the paper and that the radiation is incident on an atom at the origin O. If we had just a single electron at the origin, then the amplitude of the radiation scattered in a direction defined by the unit vector s, also in the plane of the paper, is given by electromagnetic theory as

$$E = \left(\frac{e^2}{mc^2} \right) \frac{E_0}{R}, \qquad (2.1)$$

3

Liquid Metals

when the distance R at which the amplitude is measured is large. If we had all the electrons in an atom concentrated at the origin, then obviously we would simply multiply eqn. (2.1) by the atomic number Z to find the total amplitude. But, because the electrons are distributed over regions comparable in size with the X-ray wavelength, there will be interference between the X-rays scat-

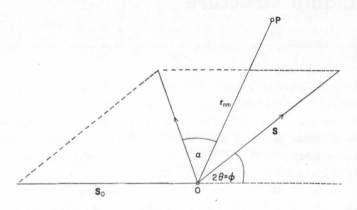

FIG. 1. Polarization in X-ray scattering

tered by different electrons in the same atom, and the atomic scattering factor f, essentially the Fourier transform of the electron density in the atom, must be introduced. In general, f is less than or equal to Z.

Now consider a second atom at position $\mathbf{r}_2$ relative to the first. The phase of the radiation scattered from the second atom compared to that scattered from the atom at the origin is $(2\pi/\lambda)\mathbf{r}_2 \cdot (\mathbf{s}_0 - \mathbf{s})$. Therefore if we write the time dependence of the incident wave as $\exp(2\pi i \nu t)$, then the scattered wave from atom 2 has the form

$$E \propto f_2 \exp\left[2\pi i \left(\nu t - \frac{\mathbf{r}_2 \cdot \{\mathbf{s}_0 - \mathbf{s}\}}{\lambda}\right)\right]. \qquad (2.2)$$

The total amplitude observed at a point distant R along $\mathbf{s}$ is therefore

$$E = \left(\frac{e^2}{mc^2}\right)\frac{E_0}{R}\sum_n f_n \exp\left[\frac{2\pi i}{\lambda}(\mathbf{s} - \mathbf{s}_0)\cdot \mathbf{r}_n\right] \qquad (2.3)$$

4

and the intensity is obtained from (2.3) as

$$|E|^2 = \left(\frac{e^2}{mc^2}\right)^2 \frac{E_0^2}{R^2} \sum_n f_n \exp\left[\frac{2\pi i}{\lambda}(\mathbf{s}-\mathbf{s_0})\cdot\mathbf{r}_n\right]$$

$$\times \sum_m f_m \exp\left[-\frac{2\pi i}{\lambda}(\mathbf{s}-\mathbf{s_0})\cdot\mathbf{r}_m\right]. \quad (2.4)$$

If we now remove the restriction to plane polarized radiation, then the usual polarization factor is introduced (see, for example, Compton and Allison, 1935, p. 116) and if $\phi(=2\theta)$ is the scattering angle (see Fig. 1) then the result is

$$|E|^2 = \frac{1}{2}(1+\cos^2\phi)\frac{E_0^2}{R^2}\left(\frac{e^2}{mc^2}\right)^2$$

$$\times \sum_n \sum_m f_n f_m \exp\left[\frac{2\pi i}{\lambda}\{\mathbf{s}-\mathbf{s_0}\}\cdot\{\mathbf{r}_n-\mathbf{r}_m\}\right]. \quad (2.5)$$

Writing the intensity in units defined by the scattering from a single electron, we find

$$I = \sum_n \sum_m f_n f_m \exp\left[iKr_{nm}\cos\alpha\right] \quad (2.6)$$

where, as indicated in Fig. 1, $r_{nm} = |\mathbf{r}_n-\mathbf{r}_m|$, $|\mathbf{s}-\mathbf{s_0}| = 2\sin\theta$, with θ as half the scattering angle, $K = \dfrac{4\pi\sin\theta}{\lambda}$ and α is the angle between $\mathbf{s}-\mathbf{s_0}$ and $\mathbf{r}_n-\mathbf{r}_m$.

As the vector $\mathbf{r}_n-\mathbf{r}_m$ takes on all orientations, we must average over α and we readily find from eqn. (2.6) that

$$I = \sum_n \sum_m f_n f_m \left\{\frac{\sin Kr_{nm}}{Kr_{nm}}\right\} \quad (2.7)$$

which is Debye's basic equation. The scattered intensity is seen from eqn. (2.7) to depend on the atomic structure factors f_n, the scattering angle, the X-ray wavelength and the interatomic distances r_{nm}.

2.2 Monatomic liquids

In the case of liquids composed of atoms of only one type, $f_n = f_m = f$ say. If we separate out from the summation in (2.7),

5

the terms with $n = m$, then, for a liquid containing N atoms, we readily obtain

$$I = Nf^2\left[1+\sum_n{}' \frac{\sin Kr_{nm}}{Kr_{nm}}\right], \tag{2.8}$$

where the summation excludes $r_{nm} = 0$.

If we now suppose that $\varrho(r)$ is the density of atoms at distance r from the atom we are 'sitting' on, then we can replace the summation in (2.8) by an integration and we obtain

$$I = Nf^2\left[1+\int_0^R 4\pi r^2\varrho(r)\frac{\sin Kr}{Kr}\,dr\right] \tag{2.9}$$

where R is the (very large) radius of the liquid sample. If we denote the average density by ϱ_0, then we may write (2.9) in the form

$$I = Nf^2\left[1+\int_0^R 4\pi r^2[\varrho(r)-\varrho_0]\frac{\sin Kr}{Kr}\,dr+\int_0^R 4\pi r^2\varrho_0\frac{\sin Kr}{Kr}\,dr\right]. \tag{2.10}$$

But the second integral in the square bracket in (2.10) evidently denotes the X-ray scattering from a uniform density and this is all concentrated in the small angle region (yielding a δ function at the origin in the limit as R tends to infinity).

2.3 Radial distribution function and structure factor

At this stage, we define a liquid structure factor $S(K)$ by the relation

$$S(K) = \frac{I}{Nf^2} \tag{2.11}$$

and then from eqn. (2.10) we find

$$S(K) = 1+\int_0^\infty 4\pi r^2[\varrho(r)-\varrho_0]\frac{\sin Kr}{Kr}\,dr, \tag{2.12}$$

where we have not considered the delta function contribution further. This then is the fundamental relation between the observed X-ray scattering intensity and the 'density' of atoms $\varrho(r)$. Following Zernike and Prins (1927), the radial distribution function $g(r)$ is defined by setting $\varrho_0 g(r) 4\pi r^2\, dr$ equal to the number of atoms in a spherical shell of radius r and thickness dr. Obviously we then find from (2.12) that

$$S(K) = 1 + \varrho_0 \int_0^\infty 4\pi r^2 [g(r)-1] \frac{\sin Kr}{Kr}\, dr \qquad (2.13)$$

or if we note that $\sin Kr/Kr$ is the s wave in the expansion of the plane wave $\exp(i\mathbf{K}\cdot\mathbf{r})$, then we have the alternative form

$$S(K) = 1 + \varrho_0 \int [g(r)-1] \exp(i\mathbf{K}\cdot\mathbf{r})\, d\mathbf{r}. \qquad (2.14)$$

Equation (2.13) would be the obvious form required, should we wish to predict the X-ray intensity from a calculated radial distribution function. As we remarked earlier, the most fruitful approach to date has been to rewrite (2.13) by inverting the Fourier transform, when we obtain

$$g(r) = 1 + \frac{1}{8\pi^3 \varrho_0} \int [S(K)-1] \exp(i\mathbf{K}\cdot\mathbf{r})\, d\mathbf{K}$$

$$= 1 + \frac{1}{2\pi^2 \varrho_0 r} \int_0^\infty [S(K)-1] K \sin Kr\, dK. \qquad (2.15)$$

Of course, the calculation of $g(r)$ from the measured $S(K)$ is subject to some difficulties and the sort of errors which can arise have been fully discussed by Paalman and Pings (1963). It will often be convenient in what follows to work in terms of the total correlation function $h(r)$ which is simply defined as $g(r)-1$.

2.4 Forces and structure

The discussion so far has concerned itself with the two-atom correlation function $g(r)$. Unfortunately, as we shall see below, when we try to relate the forces to the structure of the fluid, a three-body correlation function comes in. Thus we need to introduce such a correlation function, $n_3(\mathbf{r}_1\, \mathbf{r}_2\, \mathbf{r}_3)$ say, defined so that the probability that volume elements $d\mathbf{r}_1$, $d\mathbf{r}_2$, $d\mathbf{r}_3$ around $\mathbf{r}_1$, $\mathbf{r}_2$ and $\mathbf{r}_3$ are occupied by molecules is

$$n_3(\mathbf{r}_1\, \mathbf{r}_2\, \mathbf{r}_3)\, d\mathbf{r}_1\, d\mathbf{r}_2\, d\mathbf{r}_3$$

Liquid Metals

As has been known for a long time (cf. Green, 1952), we can obtain an exact relation for classical fluids, between the pair function $g(r)$, the triplet correlation function n_3 and the force, provided the total potential energy in the fluid can be expressed as a sum of pair potentials $\phi(r_{ij})$. This relation, which can be derived rigorously from the classical partition function, is readily obtained by the following physical argument. We first express the radial distribution function $g(r_{12})$ in the Boltzmann form

$$g(r_{12}) = \exp\left(-\frac{U(r_{12})}{kT}\right) \tag{2.16}$$

and we recognize in doing so that $U(r_{12})$ is playing the role of a potential of mean force. The total force acting on atom 1 is therefore the negative of the gradient of $U(r_{12})$ with respect to $\mathbf{r}_1$ and this can be split up into two parts, one due to the pair force between atoms 1 and 2 and the other due to the remaining atoms. Thus, remembering that the probability of finding atom 3 in the volume element $d\mathbf{r}_3$, when atoms 1 and 2 are certainly in volume tlements $d\mathbf{r}_1$ and $d\mathbf{r}_2$ around $\mathbf{r}_1$ and $\mathbf{r}_2$ is

$$\frac{n_3(\mathbf{r}_1\,\mathbf{r}_2\,\mathbf{r}_3)\,d\mathbf{r}_3}{\varrho_0^2 g(r_{12})}$$

we may write:

$$-\frac{\partial U(r_{12})}{\partial \mathbf{r}_1} = -\frac{\partial \phi(r_{12})}{\partial \mathbf{r}_1} - \int \frac{n_3(\mathbf{r}_1\,\mathbf{r}_2\,\mathbf{r}_3)}{\varrho_0^2 g(r_{12})}\frac{\partial \phi(r_{13})}{\partial \mathbf{r}_1}\,d\mathbf{r} \ . \tag{2.17}$$

This equation, as it stands, *can be shown to be* exact. But to use il to derive the forces from the structure, it will clearly be essentiat to make some assumption about n_3. This function is therefore at the very heart of the liquids problem, and we have only a rudimentary understanding of it. We shall come back to the discussion of the three-atom correlation function in Chapter 4.

2.5 Ornstein–Zernike direct correlation function

At this point we shall introduce the so-called direct correlation function, due to Ornstein and Zernike. The idea is again to split the total correlation function $h(r)$ into a pair term, and that due to the remaining atoms, following the argument used to set up (2.17). Thus, if f is the direct correlation function, we write:

$$h(r) = f(r) + \varrho_0 \int f(\mathbf{r}-\mathbf{r}') \, h(\mathbf{r}') \, d\mathbf{r}', \qquad (2.18)$$

which may be regarded as the definition of $f(r)$. Clearly if this decomposition were to have a rigorous interpretation, some three-body correlation function should have entered the definition. Since, however, we might expect simplifications when any two atoms are far apart, in the sense that the three-body correlation function can be expressed in terms of pair correlations (cf. eqn. (4.1) below), $f(r)$ may perhaps have asymptotic significance. This we examine in Chapter 4 and show to be the case (cf. Johnson, Hutchinson and March, 1964).

Let us explore a little further the relationship between the direct and total correlation functions, implied by the definition (2.18), by going into $\mathbf{K}$ space. Then, defining the Fourier transform $\tilde{f}(K)$ of the direct correlation function as

$$\tilde{f}(K) = \varrho_0 \int f(r) \exp{(i\mathbf{K} \cdot \mathbf{r})} \, d\mathbf{r}, \qquad (2.19)$$

and similarly defining $\tilde{h}(K)$, we can immediately rewrite eqn. (2.18) in the form

$$\tilde{h}(K) = \tilde{f}(K) + \tilde{f}(K) \, \tilde{h}(K)$$

or

$$\tilde{f}(K) = \frac{\tilde{h}(K)}{1 + \tilde{h}(K)}. \qquad (2.20)$$

But, by inspection of (2.14), it is clear that $\tilde{h}(K)$ is simply $S(K) - 1$, whereas from (2.20)

$$\tilde{f}(K) = \frac{S(K) - 1}{S(K)}. \qquad (2.21)$$

Inverting eqn. (2.19) and using (2.21), we have therefore the result

$$f(r) = \frac{1}{8\pi^3 \varrho_0} \int \left[\frac{S(K) - 1}{S(K)} \right] \exp{(i\mathbf{K} \cdot \mathbf{r})} \, d\mathbf{K}$$

$$= \frac{1}{2\pi^2 \varrho_0 r} \int_0^\infty \left\{ \frac{S(K) - 1}{S(K)} \right\} K \sin{Kr} \, dK. \qquad (2.22)$$

Liquid Metals

Thus from the measured structure factor $S(K)$, we can find both the radial distribution function $g(r)$ and the Ornstein–Zernike correlation function $f(r)$. Since we cannot measure the scattering through indefinitely small angles, it is a fortunate circumstance that there is a famous thermodynamic result for the long wavelength limit ($K \to 0$) of the structure factor, namely

$$S(0) = kT\varrho_0 K_T, \qquad (2.23)$$

where K_T is the isothermal compressibility. This relation will prove of central importance in the arguments to follow and therefore a derivation of it, due to Feynman and Cohen (1956), is given in Appendix 1. The reader will be well advised, however, to leave the study of this Appendix until later, as it is closely related to arguments given in Chapter 8.

2.6 Localization of direct correlation function in K space for metals

We shall proceed immediately to discuss experimental results for $S(K)$ and $\tilde{f}(K)$ for liquid metals and compare them with available data for insulating fluids. Figures 2 and 3 show some recent data obtained by neutron scattering (North, Enderby and Egelstaff, 1967, to be published; see Enderby and March, 1966a) for liquid tin at 530°K and for liquid thallium at 600°K respectively. In Fig. 4, Pb data are plotted in terms of $\tilde{f}(K)$, and for comparison, data obtained by Henshaw (1957) for A at 84°K are also shown. The two sets of data look rather similar, down to the smallest value of K shown in Fig. 4.

TABLE 1

Values of $\tilde{f}(0)$ for liquids

	A	Ne	Na	K	Hg	Ga	Pb	Sn	Bi
$\tilde{f}(0)$	−14.6	−17.6[†]	−37.5	−40.7	−134	−106[‡]	−109	−124	−99
$T°K$	84	24	371	337	323	323	600	505	544

[†] Estimated from data of Henshaw (1958).
[‡] Taking isothermal compressibility $K_T = 4.0 \times 10^{-12}$ c.g.s. units.

However, from the measured compressibilities, we can calcu-late $S(0)$ using eqn. (2.22), and hence $\tilde{f}(0)$ from eqn. (2.20). Table 1 shows data for $\tilde{f}(0)$, taken from Enderby and March (1965), for a variety of liquids. Following these authors, we now reduce the

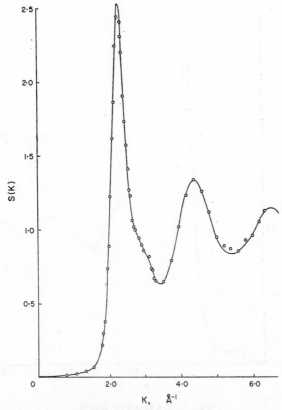

FIG. 2. $S(K)$ for Sn at 530°K

data of Fig. 4 by dividing by $\tilde{f}(0)$ and we obtain the results shown in Fig. 5. The striking difference is that $\tilde{f}(K)/\tilde{f}(0)$ is much more localized for Pb than for A. To press this point further, we show in Fig. 6 experimental results in the range $0 < K < 2\text{Å}^{-1}$ for Pb (Egelstaff, Duffill, Rainey, Enderby and North, 1966), Ga (Ascarelli, 1966) and A (Henshaw, 1957). Again, the increased localization of the ratio $\tilde{f}(K)/\tilde{f}(0)$ is clearly shown for the liquid metals.

Liquid Metals

In short, the scattering, when described by $\tilde{f}(K)/\tilde{f}(0)$ rather than by $S(K)$, shows marked localization in $\mathbf{K}$ space for metals but is much more diffuse for liquid argon. This short range nature of $\tilde{f}(K)$ is revealed particularly well if we use the data of North *et al.* for

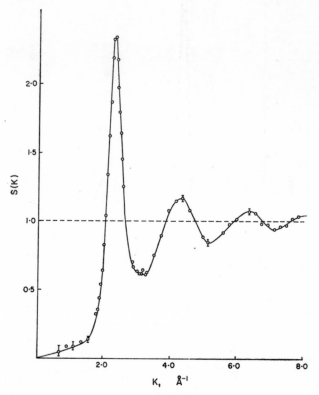

FIG. 3. $S(K)$ for Tl at 600°K

liquid thallium plotted in Fig. 3. The direct correlation function, taken from Enderby and March (1966a), is shown in Fig. 7.

Since a clear distinction exists between the measured structure data on $\tilde{f}(K)$ for liquid metals on the one hand, and liquid insulators on the other, it is natural that we turn next to consider the differences that can arise between the force laws in a liquid metal and a liquid insulator. This is the object of the following chapter.

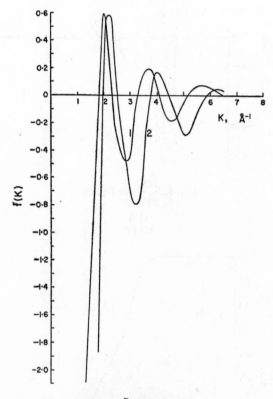

FIG. 4. $\tilde{f}(K)$ for liquids
(1) A
(2) Pb

13

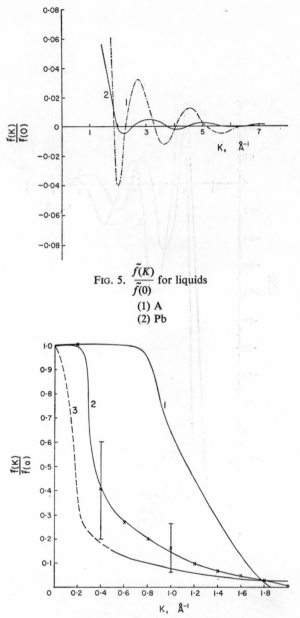

FIG. 5. $\dfrac{\tilde{f}(K)}{\tilde{f}(0)}$ for liquids

(1) A
(2) Pb

FIG. 6. Small angle scattering from (1) A; (2) Pb; (3) Ga

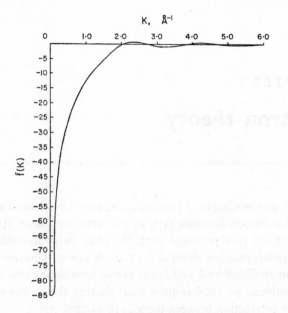

FIG. 7. $\tilde{f}(K)$ for liquid Tl

CHAPTER 3

Electron theory

WE HAVE seen in Chapter 2 how the liquid structure, defined by the radial distribution function $g(r)$, or the structure factor $S(K)$, is related to the pair potential $\phi(r)$. We shall be concerned with extracting information about ϕ from the X-ray or neutron measurements in Chapters 4 and 5, but before considering this aspect of the problem, we shall enquire what electron theory can tell us about the interaction between the ions in a liquid metal.

3.1 Pair potential for insulating fluid argon

We notice, first of all, *that interionic potentials in a liquid metal* will be less basic than those operating between the atoms of A say, in A gas, or in condensed states in A. In these cases it is generally accepted that the force law, to a first approximation, is the same in each of these three states of matter. The form is shown in curve *a* of Fig. 8. As the atoms come within a distance of the order of an atomic diameter, strong quantum-mechanical repulsive forces come into play and $\phi(r)$ rises very steeply. On the other hand, at large r, there is a relatively weak van der Waals attraction, varying as r^{-6} if we neglect retardation effects. But the situation is significantly different in a liquid metal, beyond an ionic diameter. For while it is true that, if we could expand the metal sufficiently, we should certainly reach a point at which the conduction electrons would become again attached to their own ions, and we should have a pair potential similar to curve *a* of Fig. 8 operating between neutral atoms, the force between the ions in a liquid metal is clearly dependent on the way the conduction electrons adjust themselves to shield the ions. We shall now go on to discuss the nature of this ion–ion force for point ions in a Fermi gas.

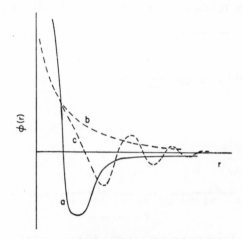

FIG. 8. Schematic forms of pair potential
(a) Between A atoms
(b) Between Pb ions in liquid Pb according to semi-classical theory
(c) According to wave theory

3.2 Fermi momentum distribution

We know from Hall constant measurements that, in a simple liquid metal like Sn, there are around four conduction electrons per atom, and if these were uniformly distributed throughout the liquid, then the density of electrons would be about $10^{23}/cm^3$. Actually, of course, the conduction electrons will not remain uniformly distributed throughout the liquid, but will pile up around the positive ions and shield them.

To see how this screening comes about, let us consider a degenerate Fermi gas of electrons, with energy states occupied up to some maximum energy E_f; the Fermi energy. For free electrons, this naturally implies that in momentum space, states are fully occupied out to a maximum momentum p_f, related to the Fermi wave number k_f by $p_f = \hbar k_f$. Thus, outside this sphere of radius p_f in momentum space, the states are totally unoccupied. This is the assumption of a sharp Fermi surface. For at least two reasons, in a liquid metal, some blurring of such a Fermi surface will occur. First, and not restricted to a liquid metal, the electrons interact

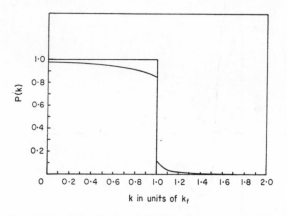

FIG. 9. Fermi distribution and effect of electron correlations

with one another, and there is a finite probability of an electron being scattered into a state outside the Fermi sphere. This obviously leads to increased kinetic energy of the electrons, but the lowering of the potential energy more than compensates for this. However, for the high electron densities appropriate to liquid metals, the work of Daniel and Vosko (1960) indicates that this effect leads to a relatively small tail on an otherwise rectangular Fermi distribution, as shown schematically in Fig. 9. Secondly, electrons will be scattered by the ionic disorder and this will lead to further Fermi surface blurring. This effect will be considered again in Chapters 7 and 9, but for the moment we shall assume the blurring to be small. This is consistent with available experimental evidence, as we shall see later.

Now, assuming a well defined maximum momentum p_f, the volume of occupied momentum space is $\frac{4}{3}\pi p_f^3$, and if Ω is the volume of the metal, then the volume of occupied phase space is $\frac{4}{3}\pi p_f^3 \Omega$. But according to the Heisenberg Uncertainty Principle, we cannot divide this phase space into cells smaller than are compatible with $\Delta p_x \Delta x \sim h$. In fact, in three dimensions, the basic cell in the phase space has a size h^3, and into it we can put just two electrons with opposed spins. Clearly then, with N electrons, we must have

$$N = \frac{2}{h^3} \cdot \frac{4\pi}{3} p_f^3 \Omega. \tag{3.1}$$

But the Fermi energy E_f for free electrons is related to p_f through

$$E_f = \frac{p_f^2}{2m} \tag{3.2}$$

where m is the electronic mass. If we write the number of electrons per unit volume N/Ω as n_0, then we have from (3.1) and (3.2)

$$n_0 = \frac{8\pi}{3h^3} (2mE_f)^{\frac{3}{2}}. \tag{3.3}$$

The above considerations correspond to the assumption that the conduction electrons are completely free. An electron with momentum $\mathbf{p} = \hbar\mathbf{k}$ has then a normalized wave function

$$\phi_\mathbf{k}(\mathbf{r}) = \Omega^{-\frac{1}{2}} \exp{(i\mathbf{k} \cdot \mathbf{r})}. \tag{3.4}$$

Such a plane wave obviously leads to a uniform probability density $\phi_\mathbf{k}^* \phi_\mathbf{k} = \Omega^{-1}$. However, when we consider the influence of a positive ion on the electrons, the wave function (3.4) is naturally changed.

3.3 Ionic shielding

Let us suppose that the ion has charge Ze, and that at first, to gain an idea of the essential features of the problem, we neglect the ion core radius, and treat Ze as a point charge. By doing so, we shall clearly destroy the validity of the force law we are trying to calculate when the atoms approach one another within distances comparable with the ionic radius. But, as we shall see, the argument is illuminating in showing us that the force law has a characteristic type of behaviour at distances considerably greater than the core radius. However, even to obtain this behaviour at

19

all quantitatively, it will be necessary to consider carefully the role of the core electrons. We shall be content here to use the over-simplified point-ion model to provide motivation for the structure data analysis of Chapters 4 and 5 and to stress the role the Fermi energy must play in determining the force law.

In a conducting medium, the electrostatic potential Ze/r created by the point ion cannot remain of such long range, since the electric field will be screened out by the polarization of the electrons by the ion. To describe this quantitatively, we must solve the Schrödinger equation for the perturbed electron states in the presence of the point ion, it being essential however, to make the calculation self-consistent in the usual Hartree sense. This implies that the perturbed wave functions calculated from the Schrödinger wave equation with a potential energy

$$V(\mathbf{r}) = \frac{-Ze^2}{r} + V_e(\mathbf{r}), \tag{3.5}$$

where V_e is the potential energy associated with the redistribution of the electrons, must reproduce this same potential.

3.3.1 SEMI-CLASSICAL THEORY

To see how this works in practice, let us make the assumption that the total potential energy $V(\mathbf{r})$ varies slowly in space. More precisely, $V(\mathbf{r})$ will be assumed to change by only a fraction of itself over a characteristic distance, which for degenerate electrons will clearly be the de Broglie wavelength for electrons at the Fermi surface, i.e. $2\pi/k_f$. Then, the constant electron density $n_0 = N/\Omega$ of eqn. (3.1) can be replaced by an approximate equation for the perturbed density $n(\mathbf{r})$ given by

$$n(\mathbf{r}) = \frac{8\pi}{3h^3} p_f^3(\mathbf{r}). \tag{3.6}$$

Naturally $n(\mathbf{r})$ now varies with the position $\mathbf{r}$ relative to the ionic centre, and, as a consequence, so must the maximum momentum $p_f(\mathbf{r})$. But the Fermi level E_f must be constant, for otherwise the electrons could readjust themselves to lower the total energy, and hence, if we write down the classical energy equation for the fastest electron, we have

$$E_f = \frac{p_f^2(\mathbf{r})}{2m} + V(\mathbf{r}). \tag{3.7}$$

Substituting for $p_f(\mathbf{r})$ in (3.6) we find

$$n(\mathbf{r}) = \frac{8\pi}{3h^3} (2m)^{\frac{3}{2}} [E_f - V(\mathbf{r})]^{\frac{3}{2}}. \tag{3.8}$$

This is the well-known semi-classical relation of Thomas and Fermi.

Thus the charge displaced around the ion, $n(\mathbf{r}) - n_0$, is given by

$$n(\mathbf{r}) - n_0 = \frac{8\pi}{3h^3} (2m)^{\frac{3}{2}} \left\{ [E_f - V(\mathbf{r})]^{\frac{3}{2}} - E_f^{\frac{3}{2}} \right\}. \tag{3.9}$$

We observe now that, without complete justification very near to the ionic charge, we can linearize (3.9) to yield

$$n(\mathbf{r}) - n_0 = -\frac{4\pi}{h^3} (2m)^{\frac{3}{2}} E_f^{\frac{1}{2}} V(\mathbf{r}). \tag{3.10}$$

But this displaced charge now screens the ion, and Poisson's equation tells us that

$$\begin{aligned}
\nabla^2 V &= 4\pi e^2 [n_0 - n(\mathbf{r})] \\
&= \frac{16\pi^2 e^2}{h^3} (2m)^{\frac{3}{2}} E_f^{\frac{1}{2}} V \\
&= q^2 V.
\end{aligned} \tag{3.11}$$

Equation (3.11) defines the quantity q, which may be written alternatively in terms of the Fermi wave number k_f as

$$q^2 = 4k_f/\pi a_0; \qquad a_0 = \hbar^2/me^2. \tag{3.12}$$

The required solution of eqn. (3.11) is evidently

$$V = -\frac{Ze^2}{r} \exp(-qr) \tag{3.13}$$

since the appropriate physical boundary conditions are that

$$V \rightarrow -Ze^2/r \quad \text{as} \quad r \rightarrow 0,$$

and that $V \rightarrow 0$ more rapidly than r^{-1} as $r \rightarrow \infty$.

Liquid Metals

3.3.2 WAVE THEORY

Equation (3.11), first given by Mott (1936) and subject to the restriction of slowly varying potentials, will now be generalized to yield the exact first-order result (cf. eqn. (3.25) below). Rather than follow the original density matrix arguments (March and Murray, 1960), we give an elementary derivation below from the Schrödinger equation (cf. March and Murray, 1961, Appendix 1). Thus if the self-consistent potential energy in which all electrons move is $V(\mathbf{r})$ as above, then the original plane waves $\Omega^{-\frac{1}{2}} \exp(i\mathbf{k} \cdot \mathbf{r})$ of the conduction electrons are distorted into wave functions $\psi_{\mathbf{k}}(\mathbf{r})$, $\mathbf{k}$ labelling the unperturbed state from which $\psi_{\mathbf{k}}$ derives, when the ion is introduced. We must then solve

$$\nabla^2 \psi_{\mathbf{k}} + \frac{2m}{\hbar^2} [E_{\mathbf{k}} - V(\mathbf{r})] \psi_{\mathbf{k}} = 0. \tag{3.14}$$

Let us work by analogy with Poisson's equation

$$\nabla^2 \phi = -4\pi\varrho. \tag{3.15}$$

A formal solution of eqn. (3.15) can be written down immediately as

$$\phi(\mathbf{r}) = \int d\mathbf{r}' \, \frac{\varrho(r')}{|\mathbf{r} - \mathbf{r}'|} \tag{3.16}$$

and clearly arises from taking the right-hand side of (3.15), multiplied by a special solution of the left-hand side equated to zero (i.e. Laplace's equation) and integrated over $\mathbf{r}'$. Obviously $1/|\mathbf{r} - \mathbf{r}'|$ is such a solution, with the property that it is singular at $\mathbf{r} = \mathbf{r}'$.[†]

Now we apply this argument to the Schrödinger equation

$$\nabla^2 \psi_{\mathbf{k}} + k^2 \psi_{\mathbf{k}} = \frac{2m}{\hbar^2} V(r) \psi_{\mathbf{k}}, \tag{3.17}$$

where we have written $E_{\mathbf{k}}$ in eqn. (3.14) explicitly as $\hbar^2 k^2 / 2m$.

Proceeding by analogy with the solution (3.16) of Poisson's equation above, we can solve eqn. (3.17) when the right-hand side

[†] It is a simple example of a Green function, which leads immediately to the solution (3.16).

equals zero and we find the spherical wave solution

$$G(\mathbf{r}, \mathbf{r}') = \frac{e^{ik\,|\mathbf{r}-\mathbf{r}'|}}{|\mathbf{r}-\mathbf{r}'|}, \tag{3.18}$$

which obviously reduces to $1/|\mathbf{r}-\mathbf{r}'|$ when $\mathbf{k} = 0$.

Then we have, for the full solution

$$\psi_{\mathbf{k}}(\mathbf{r}) = \Omega^{-\frac{1}{2}}e^{i\mathbf{k}\cdot\mathbf{r}} - \frac{m}{2\pi\hbar^2}\int d\mathbf{r}'\, G(\mathbf{r}, \mathbf{r}')\, V(\mathbf{r}')\, \psi_{\mathbf{k}}(\mathbf{r}'). \tag{3.19}$$

Evidently the first term on the right-hand side is the wave function before the potential is introduced, and the second term is obtained in exact analogy to eqn. (3.16). Now we approximate in (3.19) by noting that in the 'correction' term proportional to V, we can write $\psi_{\mathbf{k}}(\mathbf{r}') \doteq \Omega^{-\frac{1}{2}} \exp(i\mathbf{k}\cdot\mathbf{r}')$ and work simply to first order in V. To form the electron density $n(\mathbf{r})$, we must sum $\psi_{\mathbf{k}}^*(\mathbf{r})\,\psi_{\mathbf{k}}(\mathbf{r})$ over all $\mathbf{k}$ out to the Fermi surface. Remembering that, to be consistent, we must only retain terms of first order in V, we find

$$\sum_{|\mathbf{k}| < k_f} \psi_{\mathbf{k}}^*(\mathbf{r})\,\psi_{\mathbf{k}}(\mathbf{r}) = \sum_{|\mathbf{k}| < k_f} \Omega^{-1}$$

$$- \sum_{|\mathbf{k}| < k_f} \Omega^{-1}\frac{m}{2\pi\hbar^2}\int d\mathbf{r}'\, V(\mathbf{r}')\,[G(\mathbf{r}\mathbf{r}')e^{i\mathbf{k}\cdot(\mathbf{r}'-\mathbf{r})} + G^*(\mathbf{r}\mathbf{r}')e^{-i\mathbf{k}\cdot(\mathbf{r}'-\mathbf{r})}]. \tag{3.20}$$

But the summation over $\mathbf{k}$ can be replaced by an integration, and we find, remembering that there are $\Omega/(2\pi)^3$ states per unit volume of $\mathbf{k}$ space and two spin directions,

$$n(\mathbf{r}) = n_0 - \frac{2m}{(2\pi)^4\,\hbar^2}\int d\mathbf{r}'\, V(\mathbf{r}')$$

$$\times \int_{|\mathbf{k}| < k_f} d\mathbf{k}[G(\mathbf{r}, \mathbf{r}')\,e^{i\mathbf{k}\cdot(\mathbf{r}'-\mathbf{r})} + G^*(\mathbf{r}, \mathbf{r}')\,e^{-i\mathbf{k}\cdot(\mathbf{r}'-\mathbf{r})}]. \tag{3.21}$$

Since, from eqn. (3.18), G depends only on the magnitude of k, we see that integrating over the angles of $\mathbf{k}$ simply replaces $e^{\pm i\mathbf{k}\cdot(\mathbf{r}'-\mathbf{r})}$ by $\dfrac{\sin k\,|\mathbf{r}-\mathbf{r}'|}{k\,|\mathbf{r}-\mathbf{r}'|}$ as can be verified directly, or by noting that this is simply the s term in the expansion of a plane wave into spherical

23

Liquid Metals

waves. We then obtain, combining G and G^* from eqn. (3.18),

$$n(\mathbf{r}) = n_0 - \frac{2m}{(2\pi)^4 \, \hbar^2} \int d\mathbf{r}' V(\mathbf{r}')$$

$$\times \int_0^{k_f} dk 4\pi k^2 \left[\frac{\sin k \, |\mathbf{r}-\mathbf{r}'|}{k \, |\mathbf{r}-\mathbf{r}'|} \cdot \frac{2 \cos k \, |\mathbf{r}-\mathbf{r}'|}{|\mathbf{r}-\mathbf{r}'|} \right]. \quad (3.22)$$

Performing the integration over k, the final result for the displaced charge $n(\mathbf{r}) - n_0$ may be written as

$$n(\mathbf{r}) - n_0 = - \frac{mk_f^2}{2\pi^3 \hbar^2} \int d\mathbf{r}' V(\mathbf{r}') \frac{j_1(2k_f \, |\mathbf{r}-\mathbf{r}'|)}{|\mathbf{r}-\mathbf{r}'|^2} \quad (3.23)$$

where

$$j_1(x) = x^{-2}[\sin x - x \cos x],$$

the first-order spherical Bessel function. To make contact with the semiclassical theory of section 3.3.1, we observe that if V varies slowly in space, then $V(\mathbf{r}')$ may be replaced approximately by $V(\mathbf{r})$, when we find

$$n(\mathbf{r}) - n_0 = - \frac{mk_f^2}{2\pi^3 \hbar^2} V(\mathbf{r}) \int d\mathbf{r}' \frac{j_1(2k_f |\mathbf{r}-\mathbf{r}'|)}{|\mathbf{r}-\mathbf{r}'|^2}. \quad (3.24)$$

Performing the integration over $\mathbf{r}'$, we then regain eqn. (3.10). Combining the first order result (3.24) with Poisson's equation, we find the correct generalization of the Mott equation (3.11), defining the self-consistent potential energy $V(\mathbf{r})$ due to a point-ion in a Fermi gas, namely:

$$\nabla^2 V = \frac{me^2}{\hbar^2} \cdot \frac{2k_f^2}{\pi^2} \int d\mathbf{r}' V(\mathbf{r}') \frac{j_1(2k_f |\mathbf{r}-\mathbf{r}'|)}{|\mathbf{r}-\mathbf{r}'|^2}. \quad (3.25)$$

Without solving eqn. (3.25), we can see that the displaced charge can have a very different character at large r, depending on whether we use the wave theory result (3.23) or the semi-classical result (3.10). Thus, let us take a simple case when V is very short range, i.e. write $V(\mathbf{r}) = \delta(\mathbf{r})$. Then the displaced charge according to eqn. (3.10) is of similar short range while eqn. (3.23) gives, on the contrary

$$n(\mathbf{r}) - n_0 \sim \frac{j_1(2k_f r)}{r^2} \quad (3.26)$$

and thus at large r

$$n(\mathbf{r}) - n_0 \sim \frac{\cos 2k_f r}{r^3} \qquad (3.27)$$

using the form of j_1 given by eqn. (3.23).

This is then the new feature, and if we use Poisson's equation with this asymptotic form of the displaced charge, we find that the potential $V(\mathbf{r})$ also goes as $\dfrac{\cos 2k_f r}{r^3}$ for large r. That such 'wiggles' exist in the displaced charge round a given fixed perturbation in a Fermi gas was first pointed out explicitly by Blandin, Daniel and Friedel (1959).

Numerical solutions of eqn. (3.25) for $V(\mathbf{r})$ are tabulated by March and Murray (1961). This Hartree calculation, carried out self-consistently in the Born approximation, is completely equivalent to the calculation of Langer and Vosko (1959) who used many-body perturbation theory. This equivalence was first pointed out to the writer by Ostrowsky and independently by Dick (see March and Murray, 1962). The merit of the calculation of Langer and Vosko is that it affords a systematic way of estimating the correction terms to the Hartree theory in a high density gas. The present approach is adopted because of its simplicity. The numerical calculations of Langer and Vosko are also, of course, highly relevant for our purposes as they are equivalent to the results obtained by solving eqn. (3.25). The screened potentials calculated here are intimately related to the ion–ion interaction for the case of point ions as we now discuss.

3.4 Dielectric constant

As yet we have confined ourselves to a discussion of the screened potential $V(\mathbf{r})$ round a point ion. It proves useful to consider the Fourier components $\tilde{V}(\mathbf{k})$ of this potential defined by

$$\tilde{V}(\mathbf{k}) = \int d\mathbf{r} \, e^{i\mathbf{k} \cdot \mathbf{r}} V(\mathbf{r}). \qquad (3.28)$$

If we use first the screened Coulomb potential (3.13) of the semi-classical theory, then we find

$$\tilde{V}(k) = \frac{-4\pi Ze^2}{k^2+q^2}.$$ (3.29)

More interestingly, the wave theory can also be dealt with analytically in **k** space, unlike the situation in **r** space, where, as we saw above, recourse had to be made to numerical methods. By Fourier transforming eqn. (3.25) we obtain

$$\tilde{V}(k) = \frac{-4\pi Ze^2}{k^2+\dfrac{k_f}{\pi a_0}g\left(\dfrac{k}{2k_f}\right)}$$ (3.30)

where

$$g(x) = 2+\frac{x^2-1}{x}\ln\left|\frac{1-x}{1+x}\right|.$$

This important result is derived in Appendix 2. It is of some consequence that, in the long wave-length limit $k \to 0$, $g(x) \to 4$ and we find the same result

$$\tilde{V}(0) = \frac{-4\pi Ze^2}{q^2}$$ (3.31)

from both equations (3.29) and (3.30).

Often it proves valuable to express these results in terms of a dielectric constant $\varepsilon(k)$ of the Fermi gas. As is thereby implied, $\varepsilon(k)$ depends on the wave number k and may conveniently be introduced for the present purposes by writing

$$\tilde{V}(k) = \frac{-4\pi Ze^2}{k^2\varepsilon(k)}.$$ (3.32)

It follows then from eqn. (3.29) that

$$\varepsilon(k) = \frac{k^2+q^2}{k^2}$$ (3.33)

while the wave theory, from eqn. (3.30), yields

$$\varepsilon(k) = \frac{k^2+\dfrac{k_f}{\pi a_0}g\left(\dfrac{k}{2k_f}\right)}{k^2}.$$ (3.34)

This latter expression appears first to have been given by Lind-hard (1954), though it is certainly at least implicit in an earlier paper by Bardeen (1937) on electron–phonon interaction.

3.5 Electrostatic ion–ion interaction

So far, we have considered the shielding of one ion in the Fermi bath of conduction electrons. But our primary concern is with the pair interaction between ions. We can calculate this in the linear approximations (3.11) and (3.25) and in each case the essential physical result is the same. We obtain the correct interaction energy provided we regard our second ion, charge Ze, at distance r from the first ion, as sitting in the screened potential of the first ion. Then it is obvious that, for the screened Coulomb potential, the pair interaction is given by

$$\phi(r) = \frac{Z^2 e^2}{r} \exp(-qr). \tag{3.35}$$

This is represented schematically in Fig. 8, curve b, and is clearly not correct, having no minimum. It is therefore essential that we use the wave theory result, which yields for large r,

$$\phi(r) \sim \frac{\cos 2k_f r}{r^3}, \tag{3.36}$$

and the form of this pair potential is sketched in curve c of Fig. 8. The argument leading to (3.35) is given in Appendix 3. The wave theory argument paralleling this was constructed by Corless and March (1961), to whose article we refer the reader.

Alternatively, we can argue from the dielectric constant that the Coulomb interaction between two ions should be simply $\dfrac{Z^2 e^2}{r}$, modified by the dielectric constant. Or, in other words, the Fourier transform of $\phi(r)$ is given by

$$\tilde{\phi}(\mathbf{k}) = \frac{4\pi Z^2 e^2}{k^2 \varepsilon(k)} \tag{3.37}$$

which is the equivalent in $\mathbf{k}$ space of eqns. (3.35) and (3.36) when we use (3.33) and (3.34) respectively (see also Langer and Vosko,

27

1959). We plot the corresponding forms of $\tilde{V}(\mathbf{k})$ for these two cases in Fig. 10, and they are seen to be rather similar. It may therefore be surprising to the reader that they lead to such different asymptotic forms (3.35) and (3.36) in $\mathbf{r}$ space.

The difference may be traced to the fact that the dielectric constant $\varepsilon(k)$ given by (3.34) has a 'kink' in it at $k = 2k_f$ (even though to graphical accuracy, $\varepsilon(k)$ from eqn. (3.34) looks to have a con-

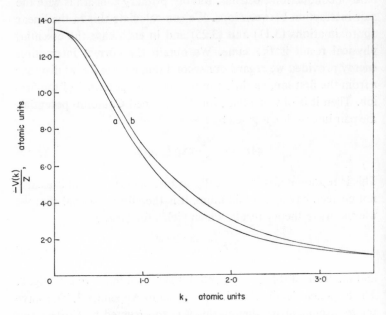

FIG. 10. Fourier components of screened point-ion interaction
(a) Semi-classical theory
(b) Wave theory

tinuous derivative), whereas (3.33) is perfectly smooth. By Fourier transform theory, the long range behaviour in $\mathbf{r}$ space is dominated either by small $\mathbf{k}$, or, if there are singularities in the $\mathbf{k}$ space form, by the singular points (cf. Lighthill, 1958). The point-ion model gives a result for the pair potential $\phi(r)$ in which all the emphasis is focused on the point $k = 2k_f$ at which $\varepsilon(k)$ is singular. But as we shall see when we turn to look at the scattering experiments in the light of this theory, the small **k** behaviour also appears to be important for finite ion cores. Only for metallic hydrogen, in the liquid state and at high density, have we any right to expect our theory to be fully quantitative!!

In conclusion then, the present screening theory indicates that the pair potential should oscillate with distance, having the asymptotic form (3.36), in contrast to the more conventional pair potential familiar for A atoms, shown in Fig. 8, curve a. We stress that the oscillations stem from the assumption of a sharp Fermi surface and would not be expected to occur if there was a blurring of around 15 to 20 per cent (see Chapter 7 below). We must now turn to enquire as to the relevance of these conclusions to the experimentally determined structure factors for liquid metals.

CHAPTER 4

Statistical theory

HAVING seen some of the unique features arising from the ionic screening by the electrons, we turn next to discuss the way in which these ideas can be combined with the existing body of knowledge on the statistical mechanics of classical fluids. For it is true that, once the electrons have been incorporated into the effective pair interaction $\phi(r)$ operating between the ions in a liquid metal, the ionic motions can be discussed classically.

In Chapter 2 we saw that a fundamental relation exists between the pair potential $\phi(r)$, the radial distribution function $g(r)$ and the three-atom correlation function n_3. Unfortunately, while $g(r)$ is accessible to experiment, n_3 is not. A problem of great current interest is just what experimental knowledge may be gained concerning n_3 (see, for example, Enderby and March, 1965, for some comments on this). But so far we have had to be satisfied with approximations to n_3, the accuracy of which is difficult to assess at the present time.

Nevertheless, we believe it true that one of the existing theories of liquids, that due to Born and Green (1946), based on the so-called superposition approximation of Kirkwood (1935) for n_3, is quite a good starting approximation for liquid metals. This is due to the long range nature of the forces discussed in the previous chapter. An alternative approach, due to Percus and Yevick (1958), which will also be considered below, seems to lead to better results for short-range forces, but these are not our main concern in this book.

4.1 Born–Green theory

We return to the basic force equation (2.17) which, as it stands, is exact within a framework in which the total potential energy function of the fluid can be expressed as a sum of pair potentials.

The idea behind the Kirkwood approximation to n_3 is to deal precisely with the correlations between pairs of atoms, but to assume the three-body correlation function can be built up as a product of pair terms. Thus, we write

$$n_3(\mathbf{r}_1\,\mathbf{r}_2\,\mathbf{r}_3) = \varrho_0^3 g(r_{12})\,g(r_{23})\,g(r_{31}) \qquad (4.1)$$

and inserting this into (2.17), we can integrate that equation with respect to $\mathbf{r}_1$. Then, as Rushbrooke (1960) has shown, the equation may readily be written in the form

$$\frac{U(r)}{kT} = \frac{\phi(r)}{kT} - \varrho_0 \int E(\mathbf{r} - \mathbf{r}')\,h(\mathbf{r}')\,d\mathbf{r}' \qquad (4.2)$$

where $h(r)$ is the total correlation function, $g(r) - 1$ and

$$E(r) = \int_r^\infty \frac{\phi'(t)}{kT}\,g(t)\,dt. \qquad (4.3)$$

Since $g(t) \to 1$ for sufficiently large t, we see from (4.3) that $E(r) \to -\phi(r)/kT$ for large r. Also we have from eqn. (2.16)

$$h(r) = \exp\left\{\frac{-U(r)}{kT}\right\} - 1, \qquad (4.4)$$

and hence the asymptotic form of $h(r)$ is given by

$$h(r) \sim \frac{-U(r)}{kT}. \qquad (4.5)$$

Bearing in mind these replacements for E and h in eqn. (4.2), and comparing it with the definition of the Ornstein–Zernike correlation function $f(r)$ in eqn. (2.18), we see a very close parallel. Whereas eqn. (4.2) tells us how to go from the potential of mean force $U(r)$ to the pair potential $\phi(r)$, the Ornstein–Zernike equation tells us how to derive a correlation function $f(r)$, intimately related to the pair interaction $\phi(r)$, from the total correlation function $h(r)$. Remembering that the parallel drawn above is only valid asymptotically $\left(\text{i.e. when } E \text{ and } h \text{ in (4.2) are replaced by } \dfrac{-\phi(r)}{kT}\right.$ and $\left.\dfrac{-U(r)}{kT} \text{ respectively}\right)$, we must expect a precise connection between $f(r)$ and $\phi(r)$ only at large r. This point is taken up again in detail in Chapter 5 (cf. also eqn. (4.7) below).

Clearly, if we knew $U(r)$, or equivalently $g(r)$, quite precisely, then we could solve eqn. (4.2) to obtain the pair potential $\phi(r)$. The results of such a procedure using the experimentally determined $g(r)$'s will be considered in the next chapter.

4.2 Hyperchain and Percus–Yevick theories

It is very desirable to have some estimate of the errors involved in the approximation (4.1) for n_3. It is fortunate therefore that we have two other theories of classical fluids, known as the hyperchain and Percus–Yevick methods. These are discussed at length by numerous authors (see the volume by Frisch and Lebowitz, 1964, or the review by Rowlinson, 1965) and we shall not therefore go into their foundations here. As the name implies, the hyperchain theory has its basis in diagrammatic analysis and while the mathematical approximations involved are clear, the physical basis of the theory remains obscure. We shall therefore confine ourselves to the remark that it is closely related to the form (4.2). We pointed out in the previous section that $f(r)$ should be rather directly connected with $\phi(r)$, and by analogy with (4.4) the roughest approximation to it might be written

$$f(r) \doteq e^{\frac{-\phi(r)}{kT}} - 1. \tag{4.6}$$

The reader familiar with the cluster expansion method of classical statistical mechanics will immediately recognize eqn. (4.6) as defining the so-called Mayer function of that theory. Equation (4.6) would then imply that for sufficiently large r

$$f(r) \sim \frac{-\phi(r)}{kT}. \tag{4.7}$$

Now $E(r)$ in (4.2), as we have seen, behaves asymptotically like $\dfrac{-\phi(r)}{kT}$, and hence, from this argument (eqn. (4.7) will be discussed more carefully below) $f(r) \doteq E(r)$ at large r. In the hyperchain theory, $f(r)$ replaces $E(r)$ everywhere, and not just asymptotically in eqn. (4.2), and we find the basic equation of that theory:

$$\frac{U(r)}{kT} = \frac{\phi(r)}{kT} - \varrho_0 \int f(\mathbf{r} - \mathbf{r}') \, h(\mathbf{r}') \, d\mathbf{r}'. \qquad (4.8)$$

This can be written in the form

$$\ln\,[1 + h(r)] + \frac{\phi(r)}{kT} = \varrho_0 \int f(\mathbf{r} - \mathbf{r}') \, h(\mathbf{r}') \, d\mathbf{r}'. \qquad (4.9)$$

The Percus–Yevick equation is closely related to (4.9). As remarked above, we shall not go into its basis, which initially stemmed from the use of collective coordinates. Suffice it to say that it has a better theoretical basis than the hyperchain equation, and it is known to give an exact solution of one problem, that of a one-dimensional assembly of hard rods. It also gives a useful description of an assembly of hard spheres, as the work of Ashcroft and March (1967) has shown. For this reason and additionally because the results are relevant to the discussion of electrical transport in Chapter 7, we summarize the exact solution of the Percus–Yevick equation for hard spheres in Appendix 4.

To write down the equation, we note that, by combining (4.9) with the definition of the Ornstein–Zernike function, we have

$$\ln\,[1 + h(r)] + \frac{\phi(r)}{kT} = h - f. \qquad (4.10)$$

The Percus–Yevick equation has a form which differs from (4.10) in that $(h - f)$ is replaced by $\ln\,(1 + h - f)$: namely

$$\ln\,(1 + h - f) = \ln\,(1 + h) + \frac{\phi(r)}{kT}. \qquad (4.11)$$

Clearly this agrees with (4.10) only when $h - f$ is sufficiently small so that $\ln\,(1 + h - f) \approx h - f$.

The investigation of Ashcroft and March (1967) referred to above points clearly to the fact that the hyperchain theory is less

33

satisfactory than either the Born–Green or the Percus–Yevick theories for hard spheres. Recent work by Gaskell (1966) also exposes serious defects in the hyperchain theory when used to calculate fluid pressure.

4.3 Force correlation function and thermodynamic consistency

Enderby and March (1966a) have pointed out that until a good theory of the three-atom correlation function is forthcoming, a useful intermediate role may be played by a new correlation function $G(r)$, defined by

$$\frac{U(r)}{kT} = \frac{\phi(r)}{kT} - \varrho_0 \int G(\mathbf{r}-\mathbf{r}')h(\mathbf{r}') \, d\mathbf{r}'. \tag{4.12}$$

This equation has the same form as those of the Born–Green and hyperchain theories, and causes one to speculate, from the fact that E and f both behave like $\dfrac{-\phi(r)}{kT}$ at large r, that the exact G has also this asymptotic form. The force correlation function $G(r)$ (strictly the correlation function for potential rather than forces) is clearly related to integrals over n_3 but as the connection does not at the moment seem fruitful, we shall not consider it further here. However, it would be interesting to calculate G, as well as n_3 of course, both by virial expansion methods and from the method of molecular dynamics (further details are given below and in Chapters 5 and 8) to see whether improvements of the various approximate theories discussed above are thereby suggested.

Due to approximations made to the three atom correlation function, or to the force correlation function $G(r)$, such as are implied when we use eqns. (4.2), (4.8) or (4.11), we cannot expect that there will be complete internal consistency when the thermodynamic predictions of the theories are considered. One useful test which is widely applied is to calculate the pressure from first the virial theorem and secondly the density fluctuation argument leading to the long wave-length limit of the structure factor $S(0)$ given by $\varrho_0 kT K_T$ (cf. Appendix 1).

These equations can be written respectively as

$$p = \varrho_0 kT - \frac{\varrho_0^2}{6} \int g(\mathbf{r})\, \mathbf{r} \cdot \frac{\partial \phi}{\partial \mathbf{r}}\, d\mathbf{r} \qquad (4.13)$$

and

$$\varrho_0 kTK_T = \varrho_0 \int [g(r) - 1]\, d\mathbf{r} + 1 \qquad (4.14)$$

where K_T is given by $\dfrac{1}{\varrho_0} \left[\dfrac{\partial \varrho_0}{\partial p} \right]_T$.

It is on the basis of these equations, into which the hyperchain equation is inserted, that Gaskell has pointed out that fluid pressures are very badly approximated by that theory.

Since departures from thermodynamic consistency are bound up with approximating to n_3 in eqn. (2.17), we shall finally consider some machine calculations of n_3 for hard spheres over a wide range of densities (Alder, 1964). The following conclusions emerge:

(i) If $n_3\,(\mathbf{r}_1\, \mathbf{r}_2\, \mathbf{r}_3)$ is calculated for $r_{12} = r_{23} = r_{31} = r$ say, then $n_3 \doteqdot \varrho_0^3 [g(r)]^3$. Indeed Alder finds that $g(r)$ derived by means of this superposition approximation from his calculated n_3 is more accurate than the best $g(r)$'s obtained theoretically so far.

(ii) At low densities, the superposition approximation is less adequate, as also shown by the later results of Ashcroft and March discussed in Appendix 4.

(iii) The superposition approximation seems likely to overestimate the net attractive forces between a pair of particles. This is because it does not take into account the shielding effects of a third particle, which would protect the pair from collisions of other particles.

It should be pointed out that conclusions drawn for hard sphere interactions are likely to overemphasize errors in the superposition approximation. The long range forces operating in liquid metals may be much more favourable for this approximation. Secondly, complete knowledge of n_3 is actually not required to relate forces and structure, any (possibly erroneous) part of n_3 which contributes zero to the integral in eqn. (2.17) being irrelevant.

35

CHAPTER 5

Pair potentials

IN CHAPTER 3 we discussed the motivation afforded by electron theory for an examination of the pair potential $\phi(r)$ in a liquid metal. The conclusion there was that, because of the diffraction of the electron waves by the ions, the potential $\phi(r)$ should have a form markedly different from that obtaining between argon atoms say. In particular, the potentials should be qualitatively different for atoms in a state which was no longer conducting and for ions, dressed with their screening clouds, in a conductor.

The argument of Chapter 3 focused attention mainly on the interaction between point ions. It turns out, however, in practice, that the structure of the ion-core plays an essential role in determining the nature of the ion–ion interaction, even at distances substantially greater than the core diameter. This electron theory problem has not been solved from first principles, though a great deal of progress has been made using the concept of pseudo-potentials (see, for example, Harrison, 1966).

We shall therefore follow here the philosophy of Johnson and March (1963; Johnson, Hutchinson and March, 1964), in which a method was developed for calculating the force law from the measured structure factors for liquids. While it has become clearer since the original work that the accuracy of the experimental data may impose substantial limitations on the accuracy of the potentials thereby obtained, there can be little doubt that a number of the essential features of the force law are established by such a procedure.

5.1 Direct correlation function and pair potential

As we pointed out in Chapter 4, the crudest approximation suggested by the Ornstein–Zernike definition (2.18) of the direct correlation function was to equate it to the Mayer function of eqn. (4.6). Clearly, we then obtain immediately a zeroth order approximation to $\phi(r)$ if we assume that $f(r)$ is calculated from the measured structure factor $S(K)$ using (2.22), from the relation

$$\phi \doteq -kT \ln \{1 + f\}. \tag{5.1}$$

This is undoubtedly too crude and in particular it fails completely when $f(r) < -1$. But it does make the point that accurate experimental determination of $f(r)$ will yield, at least in principle, information about the pair potential $\phi(r)$.

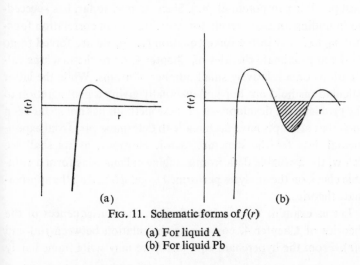

FIG. 11. Schematic forms of $f(r)$

(a) For liquid A
(b) For liquid Pb

In particular, at large r, such that $f(r) \ll 1$, (5.1) leads us to expect that $f(r)$ should be the negative of the pair potential, in units of kT. Following the proposal of Johnson, Hutchinson and March (1964) that one should examine the large r form of $f(r)$ to determine $\phi(r)$, Ascarelli (1966) and Enderby and March (1965) for liquid metals and Mikolaj and Pings (1967) for A, have developed this method further. The work of these latter authors completely confirmed the general character of $f(r)$ (cf. Fig. 11(a)) at

37

large r found by Johnson *et al.*, for A. While, for liquid metals, the long range form of the pair potentials remains a somewhat controversial matter, all workers are converging on the conclusion that there is a repulsive region in $\phi(r)$, beyond the principal minimum, for all the metals so far considered. Examples of such interactions will be considered in detail below. All the presently available evidence suggests that $f(r)$ for liquid metals has an additional minimum as in Fig. 11(b). The shaded region corresponds to the repulsive part of the potential referred to above.

5.1.1 POTENTIALS FROM HYPERCHAIN AND PERCUS-YEVICK THEORIES

From the point of view of the general theory, let us assume that we have precise knowledge of $f(r)$ and then wish to extract the best possible pair potential $\phi(r)$. Since no one, so far, has succeeded in finding an exact result for the three-atom correlation function n_3 to insert in the force equation (2.16), we are forced to go to the approximate theories of Chapter 4, or to the machine calculations on a relatively small number of atoms. While the latter calculations hold out hope of eventually giving a final solution to the problem, the development of these methods has not so far been such that attempts have been made to determine $\phi(r)$ from experimental data for the structure factor. However, as we shall see below, the available data from machine calculations form a valuable check on the analyses performed to get $\phi(r)$ using the approximate theories.

Let us examine then, a little further, the consequences of the theories of Chapter 4, concerning the relation between $f(r)$ and $\phi(r)$. From the hyperchain eqn. (4.10) we may write immediately

$$\phi_{\mathrm{hc}} = kT\{h-f- \ln [1+h]\} \qquad (5.2)$$

while for the Percus–Yevick theory defined by eqn. (4.11)

$$\phi_{\mathrm{P-Y}} = kT \ln \left[1-\frac{f}{g} \right]. \qquad (5.3)$$

As Gaskell (1966) has pointed out, if we start from the same data for $f(r)$ and $g(r)$, then the condition $\phi_{\mathrm{hc}}(r) \geqslant \phi_{\mathrm{P-Y}}(r)$ follows from (5.2) and (5.3).

Some further consequences of (5.2) and (5.3) can be seen immediately. Thus, from eqn. (5.2), we notice that at the nodes of the total correlation function $h(r)$, $f(r)$ equals $-\phi_{hc}/kT$, and also, as anticipated from the form of (5.1), this relation follows asymptotically. This latter remark needs qualification in the sense that, by expanding the right-hand side of (5.2) in powers of h, we see that this asymptotic form is true provided $h^2 \ll |f|$. This relation may be expected to hold when we are not near to the critical point (see section 5.4 below).

The Percus–Yevick theory leads also to the same asymptotic form. There are reasons, from an examination of the diagrammatic methods at large r, for thinking that this result is, in fact, correct, within the range stated above; that is far from the critical point. Unfortunately, as we discuss below, the Born–Green theory does not lead to precisely the same result, though it does again yield the important linear relation between $f(r)$ and $\phi(r)$. The proportionality constant is different, however, and this point is discussed at the end of Appendix 5. The difference can be very serious for short range forces, but should be less so for the longer range forces existing in liquid metals.

We shall use the results of this section when we come to discuss various specific metals, and their force laws. However, it does seem highly relevant to remark at this point that while it is attractive to suppose that, by making suitably accurate measurements to determine $f(r)$, we can look rather directly at the force law, there is evidence that, in one case where we can solve the Percus–Yevick equation exactly, that of hard spheres (cf. Appendix 4), the comparison of this exact solution with the correct virial expansion shows immediately that, the potential is not so directly related to the direct correlation function as the Percus–Yevick equation implies. Thus, we find that, for hard spheres, the direct correlation function cuts off discontinuously to zero at the hard core diameter according to the Percus–Yevick theory, whereas the correct virial expansion shows clearly that even for hard spheres it is non-zero outside the range of the force. This is true, in spite of the fact that the structure factor $S(K)$ appears to be quite well given by the Percus–Yevick theory in this case. Thus, while the evidence seems to suggest that there is an intimate connection between the pair potential $\phi(r)$ and $f(r)$, this connection may be overemphasized by these theories. We turn briefly then to discuss the Born–Green theory, which gives a much less direct, though still significant, relation between these two quantities.

5.1.2 POTENTIALS FROM BORN–GREEN THEORY

As remarked above, the Born–Green results for the pair potential are less easy to express directly in terms of f and g. Even the asymptotic relation (4.7) does not follow, at least in the case of classical van der Waals fluids, as discussed in Appendix 5, though $f(r)$ is again proportional to $\phi(r)$ asymptotically. We explain in that Appendix why we do not expect the differences between the asymptotic behaviours of the three theories to be so serious in liquid metals as in insulating fluids like argon. Nevertheless, attention should be drawn to the fact that the Kirkwood approximation (4.1) is not sufficiently precise to yield (4.7), which we believe to be correct when $h^2 < |f|$.

In their original investigations, Johnson and March (1963) and Johnson, Hutchinson and March (1964) worked directly from the radial distribution function $g(r)$ as tabulated in the experimental papers. Then it was not so clear that the small angle scattering was as important as we have demonstrated by examination of $\tilde{f}(K)$ in Chapter 2. It is the localization of the direct correlation function in **K** space which is leading to the long range form in **r** space for liquid metals.

Thus, more precise results than they have obtained must await detailed experimental results, preferably for a variety of temperatures. Though their work showed some oscillatory features (we shall consider their numerical results for Al and Pb presently), and although the wavelength of the oscillations was of the order of the wavelength predicted from the point ion model of Chapter 2, namely π/k_f, the present evidence (see Enderby and March, 1965) makes it unlikely that the pair potential was dominated by the region around $2k_f$, out to the largest distances at which the potentials were evaluated. Nevertheless we expect the first repulsive region in $\phi(r)$ to eventually merge with the region dominated by the sharp Fermi surface diameter $2k_f$ at sufficiently large r, though the amplitude of these latter oscillations will also be influenced by the core size.

5.2 Potentials for Al, Pb and Ga

The derived pair potentials for Al and Pb are shown in Figs. 12 and 13 respectively. As the general features are the same for both approximate eqns. (4.2) and (4.11) and since, as Johnson, Hutchinson and March emphasized, the Born–Green theory,

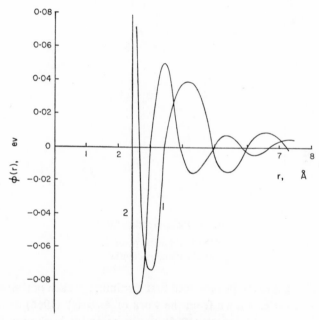

FIG. 12. Pair potentials for Al
(1) Liquid metal potential
(2) Harrison's potential

leading to relatively temperature independent potentials, seems more appropriate for liquid metals, we show only the Born–Green results. On the other hand, Johnson *et al.* concluded that for the short range forces in A, the Percus–Yevick method was more appropriate. This accords with the recent calculations of Ashcroft and March (1967) on hard sphere fluids, which are summarized in Appendix 4. These show that the Percus–Yevick method is the best of the three theories in this case.

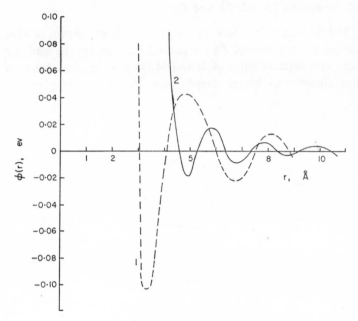

FIG. 13. Pair potentials for Pb

(1) Liquid metal potential
(2) Harrison's potential

The origin of the pronounced first maximum in each of the force curves has been shown from the work of Ascarelli (1966) on Ga, and Enderby and March (1965) on Pb, to reside in the small K behaviour of $\tilde{f}(K)$. Thus, in contrast to our point-ion model, which would suggest that all the oscillatory features come from a discontinuity in slope at $K = 2k_f$, the Fermi surface diameter, for ions with core electrons the small K behaviour also leads to similar effects, at least in the region 4–6 Å. However, Ascarelli's potential for Ga (Fig. 14), while showing a strongly repulsive region following the principal minimum, does not exhibit further well-defined oscillations[†]. We do not believe that the Percus–Yevick theory used by Ascarelli is sufficiently accurate for fluid metals, and the calculations should be repeated using the Born–Green theory. Furthermore, the method of Johnson *et al.*, starting from $g(r)$, puts in the physical condition that $g(r)$ is zero inside an atomic diameter, whereas incomplete **K** space data do not. The

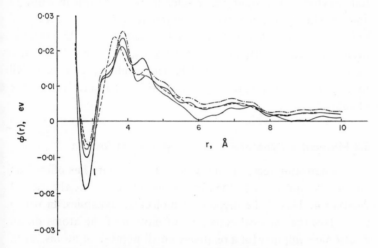

FIG. 14. Pair potentials for Ga. Derived by Ascarelli from Percus–Yevick theory. Curve 1 corresponds to $T = 150°C$. Other curves are for $T = 50°C$ and somewhat different values of K_T (between 2.4 and 4.5×10^{-12} cm²/dyn)

importance of this information needs further investigation and may in part be responsible for the difference between the **r** space and **K** space calculations. What is by now established is that the direct correlation function $f(r)$ for a liquid metal has a substantially different character from liquid Ar, as shown earlier in Fig. 11(b), where the shaded region comes in large part from the small angle scattering. It seems clear that, in spite of the theoretical interest in the long range oscillatory behaviour, precise values for the position and amplitude of the principal minimum and the

---

[†] Results resembling Ascarelli's have also been obtained for Bi and Tl (Enderby and March, 1966a) by working from the **K** space data and using the Percus–Yevick theory.

following maximum are going to be the most important parameters in the force law for many practical purposes.

The dielectric shielding of a complex ion core is a problem of great interest, and a preliminary attack on it has been made by Harrison (1963, 1964, 1965). It is then gratifying that, for Al and Pb, his calculations on the solid have subsequently led to potentials having rather similar character to those shown in Figs. 12 and 13, where his results are also plotted for comparison. It is also relevant to remark here that the ion–ion interaction in Li metal has been obtained recently by Meyer, Nestor and Young (1965), also using a pseudopotential approach. The results agree rather well in amplitude, wavelength and phase with the liquid force curve of Johnson *et al*.

5.3 Molecular dynamics and pair potential for Na

An interesting recent development in the theory has offered an alternative approach to the approximate liquid state theories of Chapter 4. This is the method of molecular dynamics, in which one solves the classical equations of motion of the atoms on an electronic computer for a relatively small number of atoms, with, say, periodic boundary conditions.

Paskin and Rahman (1966) have taken a potential near to that calculated by Johnson *et al*. for Na, by the Born–Green theory, truncating it at $r = 8.2$ Å. This potential is compared with the Born–Green potential in Fig. 15. This truncated potential was then used to calculate $g(r)$, using approximately 600 atoms in a box. The result is shown in Fig. 16, curve 1. For comparison, the X-ray data used by Johnson *et al*. are plotted in curve 2. The original papers unfortunately referred to the neutron data of Gingrich and Heaton (1961), but inspection of the temperatures used soon shows that, in fact, the X-ray data of Orton, Shaw and Williams (1960) was employed. This necessitates some changes in the conclusions of Paskin and Rahman, as pointed out below. The agreement seems good, on the whole, bearing in mind that the potentials used by Paskin and Rahman were truncated. We believe therefore that, in contrast to the conclusion drawn by Paskin and

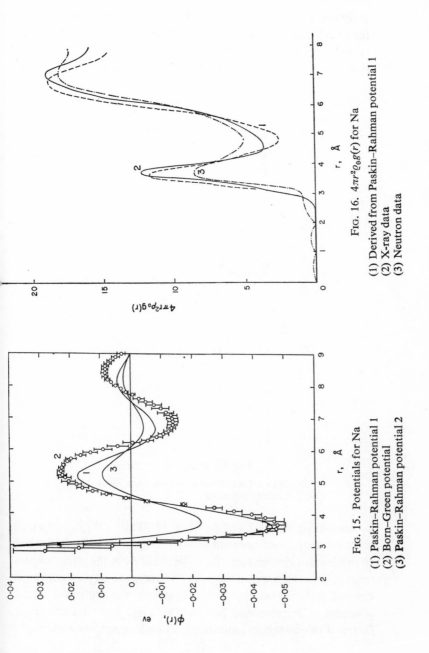

FIG. 15. Potentials for Na

(1) Paskin–Rahman potential 1
(2) Born–Green potential
(3) Paskin–Rahman potential 2

FIG. 16. $4\pi r^2 \varrho_0 g(r)$ for Na

(1) Derived from Paskin–Rahman potential 1
(2) X-ray data
(3) Neutron data

45

Liquid Metals

Rahman, the Born–Green equation is a very useful starting point for liquid metals. However, as emphasized by Enderby and March (1965), for Na major differences occur in the first and second peaks of $g(r)$ between the X-ray and neutron data. This is shown

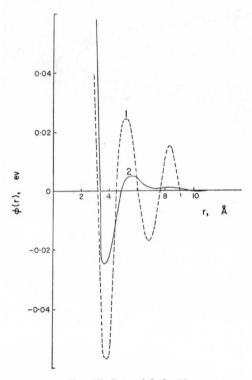

FIG. 17. Potentials for Na
(1) Born–Green potential derived from X-ray data
(2) Cochran potential

by comparing curves 2 and 3 of Fig. 16. While a third, and much more extensive, measurement of $S(K)$ is therefore essential for Na, to resolve this discrepancy, the indications from the phonon spectrum in the solid are that the neutron data are to be preferred. Paskin and Rahman showed in fact, that to fit the neutron data, the potential of Johnson and March (1963) had to be reduced by a factor of two at the first minimum. The resulting potential resembles somewhat that derived from the phonon spectrum in the solid by Cochran (1965) which is compared with the Born–Green potential derived from X-ray data in Fig. 17. The new potential is shown in curve 3 of Fig. 15, and the resulting $g(r)$ is compared with that found with potential (1), and with the neutron data, in Fig. 18.

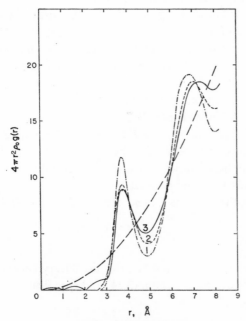

FIG. 18. $4\pi r^2 \varrho_0 g(r)$ for Na

(1) From Paskin–Rahman potential 1
(2) From Paskin–Rahman potential 2
(3) Neutron data

5.4 Asymptotic form of radial distribution function

We shall now turn to discuss briefly the asymptotic behaviour of the radial distribution function $g(r)$ in a liquid metal. According to the classical theory of Ornstein and Zernike, $g(r)$ should approach the asymptotic limit of unity in exponential fashion. However, as has been emphasized in the work of Enderby et al. (1965), this result follows only if $S(K)$ is analytic in K. This appears not to be the situation either in liquid metals, or in fact in a liquid insulator like Ar. We stress that our arguments are only valid when we are well away from transition points.

The theory hinges on the relation $f(r) \sim -\phi(r)/(kT)$ discussed earlier. One qualification should be made at this point. While all the approximate theories yield $f \alpha \phi$ at large r, the Born–Green theory is now known to give a different proportionality constant than $(-1/kT)$ for van der Waals liquids (Gaskell, 1965, see also Appendix 5). However, we shall proceed here on the assumption, remarked on earlier, that the Percus–Yevick and hyperchain methods are leading to the correct asymptotic form.

Returning to the analytic behaviour of $S(K)$, it emerges that, whereas in A the van der Waals force Ar^{-6} requires the presence of a K^3 term in the small K expansion of $S(K)$ (see Appendix 5 for a detailed argument), thereby making $S(K)$ non-analytic at the origin, in liquid metals the dielectric screening theory of section 3.4 yields a singular point at $K = 2k_f$.

Assuming that at large r, in accordance with eqn. (3.36),

$$\phi(r) \sim \frac{B \cos (2k_f r)}{r^3}, \tag{5.4}$$

then, as discussed in section 3.5, this oscillatory behaviour arises from the **K** space form of eqn. (3.34) and specifically from a term behaving, near $K = 2k_f$, as

$$(K-2k_f) \ln |K-2k_f|. \tag{5.5}$$

Using the asymptotic relation $f(r) \sim -\phi(r)/(kT)$, and that $\tilde{f}(K)$ and $S(K)$ are related by eqn. (2.21) it follows that $S(K)$ has the following expansion around $2k_f$:

$$S(K) = S(2k_f) + S_1, \tag{5.6}$$

where S_1, related directly to eqn. (5.5), dominates the asymptotic behaviour in **r** space. $h(r)$ is then immediately related to $f(r)$ since from eqn. (2.20)

$$\frac{S-1}{S} \sim \text{const.} + \frac{S_1}{\{S(2k_f)\}^2} \tag{5.7}$$

and the Fourier transform of $h(r)$ is given by

$$S-1 \sim \text{const.} + S_1. \tag{5.8}$$

Since S_1, as we have seen, leads to the asymptotic form $r^{-3} \cos 2k_f r$, we may evidently write from eqns. (5.7) and (5.8)

$$f(r) \sim -\frac{B \cos 2k_f r}{kTr^3} \tag{5.9}$$

and

$$h(r) \sim -\frac{B\{S(2k_f)\}^2 \cos 2k_f r}{kTr^3}. \tag{5.10}$$

Since $S(2k_f)$, from experimental structure data on liquid metals, is always of the order of unity, we expect $h(r)$ and $f(r)$ to have approximately the same amplitudes at sufficiently large r, in complete contrast to the situation discussed in Appendix 5 for liquid insulators.

We should stress finally that some modification of this simple model will occur at sufficiently large r, owing to the damping of the $r^{-3} \cos 2k_f r$ behaviour by slight Fermi surface blurring (Gaskell and March, 1963; see also Chapter 7 below).

However, we do not expect this to modify our essential conclusion that, for liquid metals, $h(r)$ and $f(r)$ both oscillate about zero asymptotically. Nevertheless, if the vertical discontinuity in the Fermi momentum distribution is removed by the disorder scattering, as we would expect, then eventually exponential damping of the form $e^{-\Delta k \cdot r}$ will occur, where Δk gives a measure of the Fermi surface blurring. The oscillatory behaviour of $h(r)$ and $f(r)$ which we expect for liquid metals for very large r is in marked contrast to that of classical insulating fluids like argon where, as shown in Appendix 5, $h(r)$ and $f(r)$ are expected to approach zero as an inverse power of r.

CHAPTER 6

Melting

WHILE the calculations reported in Chapter 5, using approximate theories plus structure data, seem to us still to be the best calculations we have available, we have seen that limitations in the data, plus the approximations made for n_3, leave considerable uncertainties in $\phi(r)$. Beyond the fact that $\phi(r)$ has a significant repulsive region out beyond the first minimum, as first demonstrated by Johnson and March, and confirmed subsequently by Cochran for Na, by Ascarelli for Ga and by Cowley, Woods and Dolling (1966) for K, the long-range details of the pair potential remain a controversial matter.

We wish therefore to consider some further evidence which we believe is highly relevant to screening theory and hence to the existence of long-range forces in solid and liquid metals. The essential argument is that if, as is obvious in the present picture, the Fermi energy plays an essential role in determining the forces, then presumably some correlation between melting and the Fermi energy should manifest itself. The following argument, though restricted quantitatively by the use of perturbation theory, seems to have sufficient contact with experiment to convince one that it must contain some of the essential physics of the problem.

6.1 Mukherjee's relation

The starting point of the present argument is an empirical relation due to Mukherjee (1965) between the Debye temperature θ and the energy E_v required to form a vacancy in close-packed metals in the solid state. That θ, which gives us a gross measure of the long wave limit of the phonon spectrum, should depend on the interionic forces is quite clear. What is less clear is that it will be intimately related to the energy required to remove an atom from inside the metal and place it on the surface. However, such an intimate relation does follow from known consequences of dielectric screening theory as the present writer has shown (March, 1966).

Mukherjee's relation may be written in the form

$$\theta = \frac{CE_v^{\frac{1}{2}}}{\mathcal{O}^{\frac{1}{3}} M^{\frac{1}{2}}} \tag{6.1}$$

where $\mathcal{O}$ is the atomic volume, M is the ionic mass, and C is a constant, which Mukherjee obtained from the experimental values of θ and E_v, using low temperature specific heat data to evaluate θ. We shall now outline the electron theory which leads to eqn. (6.1).

6.1.1 VACANCY FORMATION ENERGY

Let us begin by calculating the change in the one-electron energy levels when we remove an atom. In the model in which we place a negative charge $-Ze$ at the vacancy site, the perturbing potential energy is simply that calculated in section 3.3 round an ion, with a sign change. Each level is raised in first-order perturbation theory, by an amount

$$\Delta E = \frac{1}{\Omega} \int d\mathbf{r} \, e^{-i\mathbf{k}\cdot\mathbf{r}} V(r) \, e^{i\mathbf{k}\cdot\mathbf{r}}, \tag{6.2}$$

where Ω is the total volume of the metal. But this result is simply related to the long wave limit of the Fourier transform $\tilde{V}(\mathbf{k})$ of the potential calculated in section 3.4. Each level is shifted by the same amount, from eqn. (6.2), and summing over N electrons we have

$$\sum \Delta E = \frac{N}{\Omega} \pi^2 Z e^2 a_0 / (k_f)$$

$$= \frac{2}{3} Z E_f \tag{6.3}$$

Liquid Metals

from eqns. (3.6), (3.12) and (3.31). Essentially this result was obtained by Fumi (1955), even though he did not apply dielectric screening theory directly. If we follow Fumi, and neglect relaxation of the atoms into the vacancy, we have an expansion of the metal by one atomic volume when we place the atom on the surface. This increase in the volume which the conduction electrons can explore lowers the kinetic energy by (2/5) ZE_f and yields as our approximation to E_{fv}:

$$E_v = \left[\tfrac{2}{3} - \tfrac{2}{5} \right] ZE_f = \tfrac{4}{15} ZE_f. \tag{6.4}$$

We should stress that this argument is limited to small Z. Higher terms must certainly enter for Z's greater than unity, and even for monovalent metals the correction terms are quantitatively significant.

6.1.2 Velocity of sound in metals

We turn next to deal with the Debye temperature. Making the relatively rough approximation that transverse and longitudinal waves have a common velocity v_s, we have, from elementary Debye theory (cf. Mott and Jones, 1936)

$$\theta = \frac{v_s}{\varrho^{\frac{1}{3}}} \left(\frac{3}{4\pi} \right)^{\frac{1}{3}} \frac{h}{k_B}.^{\dagger} \tag{6.5}$$

We then need to obtain an approximate formula for the velocity of sound, and this has been done by Bardeen and Pines (1955; see also Bohm and Staver, 1952). Let us argue that we can treat the positive ions as though they were executing plasma oscillations. Then, since the charge on the ions is Ze and if, as usual, we write the density of ions as ϱ_0, the elementary formula for the plasma frequency ω_p is (cf. Raimes, 1961)

$$\omega_p = \left\{ \frac{4\pi\varrho_0(Ze)^2}{M} \right\}^{\frac{1}{2}}, \tag{6.6}$$

where M is the ionic mass.

---

† Here and subsequently we shall write k_B for Boltzmann's constant if there is danger of confusion with the wave number k.

Now for a metal of valency Z, we have that the mean electron density n_0 is $Z\varrho_0$, and hence eqn. (6.6) may be written

$$\omega_p = \left\{ \frac{4\pi n_0 Z e^2}{M} \right\}^{\frac{1}{2}}. \tag{6.7}$$

But we have neglected the shielding of the ions by the electrons. In the long wave limit, we can argue from either eqn. (3.29) or eqn. (3.30) that the ionic charge Ze must be screened in such a way that

$$\frac{Ze}{k^2} \to \frac{Ze}{k^2+q^2}, \quad \text{or} \quad Ze \to \frac{Zek^2}{q^2},$$

where q is the screening radius and k is the wave number. Making this replacement in eqn. (6.7) we find, as we could have anticipated, that for long wavelength phonons, the phonon frequency ω is proportional to k. If we eliminate n_0 and q in favour of the Fermi velocity $v_f = p_f/m$, then we find

$$\omega = v_s k = \left(\frac{Zm}{3M} \right)^{\frac{1}{2}} v_f k. \tag{6.8}$$

From eqn. (6.8) we may write

$$Mv_s^2 = \left(\frac{Zm}{3} \right) v_f^2$$

$$= \frac{2}{3} Z E_f. \tag{6.9}$$

Eliminating ZE_f between eqns. (6.4) and (6.9), and using eqn. (6.5) we find Mukherjee's relation (6.1). The constant C would be correctly given by the theory if, for example, $(4/15)ZE_f$ in eqn. (6.4) were replaced by $(1/6)ZE_f$. In view of the obvious approximations involved, the agreement must be regarded as quite satisfactory. We consider Mukherjee's relation to be strong support for dielectric screening theory, though a more general proof than the present first-order treatment is still lacking.

6.2 Order–disorder theory of melting

Mukherjee pointed out that the relation between θ and E_v was of the same form as the Lindemann melting point formula, which is usually written as

$$\theta = D \left(\frac{T_m}{A V^{\frac{2}{3}}} \right)^{\frac{1}{2}}, \tag{6.10}$$

where A is the atomic weight and V is the molar volume. The coefficient D is found empirically to be approximately $120 \text{ cm g}^{\frac{1}{2}}$ $\deg \text{K}^{\frac{1}{2}}$. We turn now to consider how this relation can be understood from the standpoint of electron theory.

Our thesis that long range forces exist between the ions in metals suggests immediately that the starting approximation should be that of a molecular field treatment. The early work of Lennard-Jones and Devonshire (1939; see also Frank, 1939), where the application was however to shorter range forces, and therefore open to objection, viewed melting as an order–disorder transition, and used the method of Bragg and Williams (1934).

We can best summarize the essential content of this theory as follows (Enderby and March, 1966b). We define an order parameter R, which, following Frank (1939), we can take simply as the ratio:

$$\frac{\text{Number of atoms on lattice sites}}{\text{Total number of atoms}}.$$

Then, the basic energy in the theory is that required to take an atom from an ordered to a disordered site, say U, and in the approach of Bragg and Williams, this is proportional to the order parameter. Thus we write

$$U = U_0 R, \tag{6.11}$$

and clearly U_0, like the vacancy formation energy, is an energy characteristic of a perfectly ordered crystal. Obviously, since U_0 is the only characteristic energy in the theory, the energy associated with the melting temperature T_m is of the order of U_0. Rough estimates, no better than order of magnitude, suggest that

$$k_B T_m \doteq \frac{U_0}{4}. \tag{6.12}$$

Now, while U_0 is not the vacancy formation energy, it is well known empirically that another characteristic energy Q, the activation energy for self-diffusion, is simply related to E_v (Simmons and Baluffi, 1960) by

$$E_v = 0.55Q, \tag{6.13}$$

for face-centred-cubic metals. This all suggests strongly then that $E_v \alpha U_0$, the proportionality constant being of order unity. Hence,

$$k_B T_m \sim \frac{E_v}{4}, \tag{6.14}$$

which we choose to write, using eqn. (6.4) in the form

$$k_B T_m = \beta Z E_f, \tag{6.15}$$

where β is $\sim \frac{1}{15}$. Collecting together these results, we find

$$\theta = \frac{h}{k_B} \left(\frac{3}{4\pi \mho} \right)^{\frac{1}{3}} \left(\frac{2k_B T_m}{3m\beta} \right)^{\frac{1}{2}} = D \left(\frac{T_m}{A V^{\frac{2}{3}}} \right)^{\frac{1}{2}} \tag{6.16}$$

where

$$D = \frac{h}{k_B} \left(\frac{3}{4\pi} \right)^{\frac{1}{3}} \left(\frac{2k_B}{3\beta} \right)^{\frac{1}{2}} N_a^{\frac{5}{6}}, \tag{6.17}$$

and N_a is Avogadro's number. This is of course the Lindemann relation. Although derived here from a metal model, the relation has much wider validity.

6.3 Comparison with experiment

We can now ask how eqns. (6.4) and (6.15) are related to experiment. Following Enderby and March (1966b), we plot E_v/ZE_f and $k_B T_m/ZE_f$ against Z in Figs. 19(a) and 19(b) respectively, using experimental values for E_v and T_m, and the free electron value for E_f. Extrapolating back to $Z = 0$, we find that E_v/ZE_f passes through the theoretical limit 4/15 to good accuracy. Adopting the

Liquid Metals

same procedure for $k_B T_m/ZE_f$, we find the limiting value of $\sim 1/30$, which compares reasonably with the rough estimate for β in eqn. (6.15). If we now employ this value $\beta = 1/30$ to estimate Lindemann's constant, then it follows that $D \sim 100$ cm $g^{\frac{1}{2}}$ deg $K^{\frac{1}{2}}$, whereas experiment gives 120. This again we regard as very satisfactory.

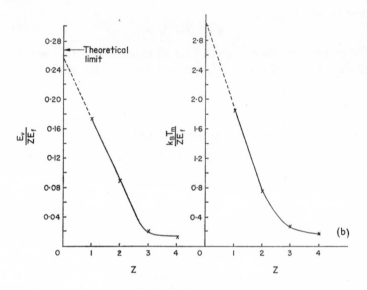

FIG. 19. (a) Vacancy formation energies in units of ZE_f, versus Z. (b) Energy $(k_B T_m)$ associated with melting temperature T_m, versus Z

Finally, if we recall the compressibility formula (2.23), then expressing it in terms of the velocity of sound, and evaluating $S(0)$ at the melting point T_m, the simple theory presented here indicates that $S_{T_m}(0)$ should be roughly constant for liquid metals (Enderby and March, 1966b). The available experimental results are shown in Table 2, taken from the above paper. This prediction is roughly borne out. It is interesting that there appears some evidence, not conclusive but suggestive, that there is a systematic difference between metals that are close-packed in solid, and the body-centred cubic alkalis. In this connection a recent investi-

Table 2

Values of S(0) at melting point for liquid metals

Metal	Cu	Ag	Mg	Al	Pb	Na	K
Z	1	1	2	3	4	1	1
T_m(°K)	1356	1233.6	924	933	600.5	371	337
$S(0)$	0.016	0.012	0.019	0.016	0.009	0.027	0.024

gation by Vosko, Taylor and Keech (1965) on phonons in solid Na, Al and Pb prompts the suggestion that the electronic screening is significantly different in body-centred and face-centred metals.

CHAPTER 7

Electrical transport

WE HAVE been primarily concerned, so far, with the structure of the liquid state and the forces between the ions. Nevertheless, we have already seen in Chapter 2 that the electrons must play a major role in determining the interionic forces and this is further confirmed by the relation between melting, valency and Fermi energy discussed in the previous Chapter.

At this point, we turn to enquire whether the scattering of the electrons off the ions can be understood, in order that we can construct a theory of the electrical resistivity of liquid metals. Early important work in the search for such a theory was that of Krishnan and Bhatia (1945), but recently the work of Ziman (1961) and his collaborators has added very substantially to our understanding of transport phenomena in liquid metals. The basic idea is that we can treat the electron wave functions as plane waves and consider them as being scattered by weak scattering potentials, arranged, of course, in a manner consistent with the measured distribution function.

It is still questionable whether such a treatment can be made fully quantitative. Thus, Greenfield (1966), on the basis of his new measurements of the structure factor $S(K)$ for liquid Na, has claimed that no choice of the single centre scattering potential can lead to agreement with the measured temperature dependence of the electrical resistivity of Na. Nevertheless, there can be little doubt that the present theory exposes many of the essential features of the problem.

We shall consider first the scattering of the electron waves by a single ionic centre in some detail and then it will be a relatively straightforward matter to incorporate later explicit accounts of the interference effects associated with the introduction of the structure factor $S(K)$.

7.1 Probability of scattering by a single ion

Suppose a single ionic centre scatters an electron from state **k** to **p**. We shall deal only with the case when the scattering is elastic, that is the electrons can only make transitions to states of the same energy.

The calculation can be carried through rather simply by means of Dirac time-dependent perturbation theory. Thus we may write the time-dependent Schrödinger equation in the form

$$ih\frac{\partial\Psi}{\partial t} = H_0\Psi + V\Psi \tag{7.1}$$

where H_0 is the Hamiltonian in the absence of the scattering centre. Let the stationary states of H_0 be $\psi_p(\mathbf{r})$ with corresponding eigenvalues E_p. Then we follow Dirac and write for the wave function at time t

$$\Psi = \sum_p a_p(t)\,\psi_p(\mathbf{r})e^{-\frac{iE_p t}{\hbar}} \tag{7.2}$$

where obviously $|a_p(t)|^2$ is the probability of the electron being found in state **p** at time t. If the electron is in state **k** at time $t = 0$ then evidently

$$\left.\begin{array}{l} a_k(0) = 1 \\ a_p(0) = 0 \quad \mathbf{p} \neq \mathbf{k}. \end{array}\right\} \tag{7.3}$$

If we substitute this into the wave equation, neglect terms involving $Va_p(\mathbf{p} \neq \mathbf{k})$, and integrate over the volume Ω of the metal after multiplying by $\psi_p^* e^{\frac{iE_p t}{\hbar}}$, we find

$$ih\frac{da_p}{dt} = V_{k\,p}e^{i(E_p - E_k)t/\hbar} \tag{7.4}$$

where

$$V_{kp} = \int \psi_p^* V\psi_k\, d\mathbf{r}. \tag{7.5}$$

Liquid Metals

If we now integrate with respect to time, employing the initial conditions, then we find

$$a_p(t) = \frac{1}{i\hbar} \frac{e^{i\,\Delta\omega t} - 1}{i\,\Delta\omega} V_{kp}, \tag{7.6}$$

where

$$\Delta\omega = \frac{E_p - E_k}{\hbar}. \tag{7.7}$$

Thus the probability that after time t the electron will be found in state p is given by

$$|a_p(t)|^2 = \frac{1}{\hbar^2} |V_{kp}|^2 \frac{2(1 - \cos \Delta\omega t)}{(\Delta\omega)^2}. \tag{7.8}$$

The time-dependent function for long times has a very strong maximum at $\Delta\omega = 0$, which is, of course, the energy conservation condition. In our case we must consider that the energies E_k lie in a continuous range, and we must integrate the transition probability over a large number of final states, for all of which p has nearly the same value.

We then require the probability per unit time, $P(k, p)\,dS$ say, of a transition to a state lying in an area dS of the Fermi distribution. To obtain this, choose a small volume $dS\,d\xi$, in which the number of states is $\frac{\Omega}{8\pi^3}\,dS\,d\xi$. The probability that after a time t the electron is in a state within it, is obtained by multiplying this by $|a_p|^2$. Then we must integrate this across the surface of the Fermi distribution, when we obtain

$$P(k, p)\,dS = \frac{\Omega\,dS}{8\pi^3\hbar^2} \frac{\partial}{\partial t} \int |V_{kp}|^2 \frac{2(1 - \cos \Delta\omega t)}{(\Delta\omega)^2}\,d\xi. \tag{7.9}$$

For points near to the Fermi surface, we may write

$$\Delta\omega = \frac{1}{\hbar} \frac{\partial E}{\partial k_n} \xi \tag{7.10}$$

where k_n is normal to the surface. Taking t long compared with $\hbar/E$ we find that the integrand has a huge maximum in the neighbourhood of $\xi = 0$ and we obtain the result

$$P(k, p)\,dS = \frac{\Omega\,dS}{4\pi^2\hbar} |V_{kp}|^2 \left/ \left| \frac{\partial E}{\partial k_n} \right| \right. . \tag{7.11}$$

We see that, apart from the matrix element $|V_{kp}|^2$, the transition probability is proportional to

$$\frac{dS}{\left|\dfrac{\partial E}{\partial k_n}\right|}$$

and thus, from band theory, to the density of states at the Fermi surface (cf. however, the remarks in §7.10 below).

If we specialize to free electrons, then we obtain immediately the probability per unit time that an electron is deflected through an angle θ into a solid angle $d\omega$. Thus

$$dS = k^2\, d\omega \tag{7.12}$$

and

$$\mathbf{k} = \frac{m\mathbf{v}}{\hbar} \tag{7.13}$$

where $\mathbf{v}$ is the velocity. We then find

$$P\, dS = \frac{v}{\Omega}\left|\frac{2\pi m}{h^2}\int e^{i(\mathbf{k}-\mathbf{p})\cdot\mathbf{r}}V(\mathbf{r})\, d\mathbf{r}\right|^2 d\omega. \tag{7.14}$$

The scattering cross-section, or the effective area that an electron must hit if it is to be scattered into a solid angle $d\omega$, is given by

$$\frac{P\Omega}{v}\, dS = \left|\frac{2\pi m}{h^2}\int e^{i(\mathbf{k}-\mathbf{p})\cdot\mathbf{r}}V(\mathbf{r})\, d\mathbf{r}\right|^2 d\omega. \tag{7.15}$$

This is the usual Born approximation formula.

7.2 Distribution function in electric field

We wish now to calculate the conductivity in terms of this transition probability (cf. Mott and Jones, 1936).

When we switch on an external electric field F, we change the Fermi distribution from its usual form

$$f_0 = \frac{1}{e^{(E-\zeta)/k_B T}+1}, \tag{7.16}$$

because the component k_z of the wave vector of an electron increases according to Newton's law (F being along the z axis)

$$\frac{dk_z}{dt} = \frac{e}{\hbar}F. \tag{7.17}$$

Liquid Metals

Thus the whole Fermi distribution is displaced as shown in Fig. 20 and has at time t the form

$$f(\mathbf{k}) = f_0 \left(k_x, k_y, k_z - \frac{eFt}{\hbar} \right). \tag{7.18}$$

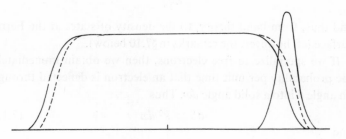

FIG. 20. Displacement of Fermi distribution in presence of electric field. Solid line is f_0, dotted line is f, and peak on right shows gradient of f_0

We are interested in the rate of change of $f(k)$ with time and if we differentiate we obtain, for $t \to 0$

$$\left(\frac{df}{dt} \right)_{\text{field}} = - \frac{\partial f_0}{\partial k_z} \frac{eF}{\hbar}$$

$$= - \frac{df_0}{dE} \cdot \frac{dE}{dk} \cdot \frac{k_z}{k} \cdot \frac{eF}{\hbar}. \tag{7.19}$$

But $\dfrac{df_0}{dE}$ is only large near the Fermi surface and only those states which lie near this surface are at first altered by the field (cf. Fig. 20).

In a liquid metal (or solid at a finite temperature), a steady state will eventually be achieved in which the effect of the field is just balanced by the collision processes considered. When this situation is reached, the distribution function is given by (7.18) with

$$t = \tau,$$

where τ is the relaxation time.

We may then write, for the perturbed distribution function

$$f(k) = f_0(k) - \frac{df_0}{dk} \cdot \frac{k_z}{k} \cdot \frac{eF}{\hbar} \, \tau. \tag{7.20}$$

7.3 Relaxation time

We have finally then to obtain an expression for the time rate of change of f due to collisions, in order to determine τ. Considering the volume $d\mathbf{k}$, the number of electrons leaving will be

$$\frac{2d\mathbf{k}}{(2\pi)^3} f(\mathbf{k}) \int [1-f(\mathbf{p})]\, P(\mathbf{k}, \mathbf{p})\, dS_\mathbf{p} \qquad (7.21)$$

where the first term is the number of electrons initially in $d\mathbf{k}$, while $(1-f(\mathbf{p}))$ is the probability that state $\mathbf{p}$ is unoccupied.

But the number of electrons entering this volume is given by

$$\frac{2d\mathbf{k}}{(2\pi)^3} [1-f(\mathbf{k})] \int f(\mathbf{p})\, P(\mathbf{p}, \mathbf{k})\, dS_\mathbf{p}, \qquad (7.22)$$

and hence the rate of change of f due to collisions is

$$\left(\frac{df}{dt}\right)_{\text{collision}} = [1-f(\mathbf{k})] \int f(\mathbf{p})P(\mathbf{p},\mathbf{k})\, dS_p - f(\mathbf{k}) \int [1-f(\mathbf{p})]P(\mathbf{k},\mathbf{p})\, dS_p. \qquad (7.23)$$

If $\left| \dfrac{dE}{dk_n} \right|$ is constant on the Fermi surface as we will assume, then

$$P(\mathbf{k}, \mathbf{p}) = P(\mathbf{p}, \mathbf{k}) \qquad (7.24)$$

and hence

$$\left(\frac{df}{dt}\right)_{\text{collision}} = \int f(\mathbf{p})\, P(\mathbf{p}, \mathbf{k})\, dS_p - f(\mathbf{k}) \int P(\mathbf{k}, \mathbf{p})\, dS_p, \qquad (7.25)$$

which would have been obtained by an elementary calculation if we had taken no account of the Pauli Exclusion Principle.
Setting

$$\left(\frac{df}{dt}\right)_{\text{collision}} + \left(\frac{df}{dt}\right)_{\text{field}} = 0, \qquad (7.26)$$

we obtain almost immediately

$$-\tau \int \left(\frac{p_z}{k_z}-1\right) P(\mathbf{k}, \mathbf{p})\, dS_p = 1. \qquad (7.27)$$

We can now rewrite this solely in terms of the angle θ between $\mathbf{k}$ and $\mathbf{p}$ and we obtain almost immediately

$$\frac{1}{\tau} = 2\pi k^2 \int_0^\pi (1 - \cos\theta)\, P(\theta) \sin\theta\, d\theta. \qquad (7.28)$$

Liquid Metals

Thus, we have to weight the differential scattering cross-section with a factor which measures the relative change in the component of the particle velocity along the initial direction of motion.

7.4 Calculation of current

Introducing $P(\theta)$ explicitly we have

$$\frac{1}{\tau} = \frac{\Omega}{2\pi h} \frac{1}{\frac{dE}{dk}} \int |V_{kp}|^2 (1 - \cos \theta) \, dS. \qquad (7.29)$$

The calculation of the current is now elementary. The current per electron is ev_z and the current density j is clearly given by

$$j = \int d\mathbf{k} \, ev_z \frac{2f(\mathbf{k})}{(2\pi)^3}. \qquad (7.30)$$

The equilibrium value of $f(\mathbf{k})$ makes no contribution to the integral and we find

$$j = -\frac{e^2 F}{\hbar^2} \frac{2}{(2\pi)^3} \int d\mathbf{k}\tau \frac{dE}{dk} \left(\frac{k_z}{k}\right)^2 \frac{df_0}{dk}. \qquad (7.31)$$

Integrating over angles we obtain

$$j = -\frac{e^2 F}{3\pi^2 \hbar^2} \int_0^\infty \frac{dE}{dk} \tau k^2 \frac{df_0}{dk} \, dk. \qquad (7.32)$$

Since $\dfrac{df_0}{dk}$ is only non-zero over a small range of k at the Fermi surface, and recalling that

$$\int \frac{df_0}{dk} \, dk = -1, \qquad (7.33)$$

we find

$$\sigma = \frac{j}{F} = \frac{e^2}{3\pi^2 \hbar^2} \left[\tau k^2 \frac{dE}{dk} \right]_{k=k_f} \qquad (7.34)$$

and the conductivity depends on the relaxation time evaluated on the Fermi surface.

64

For free electrons, we have finally, from equations (7.29) and (7.34),

$$\varrho = \frac{1}{\sigma} = \frac{3\pi^2 m}{e^2 k_f^3 \tau(k_f)}$$

$$= \frac{3m^2\Omega}{4e^2\hbar^3 k_f^2} \int |V_{kp}|^2 (1 - \cos\theta)\, d\omega. \qquad (7.35)$$

7.5 Ion–ion correlations and electron scattering

We have reached the point at which we must now introduce the fact that the ions in the liquid, with positions $\mathbf{R}_i$, are far from independent. This will then lead us directly to an expression for $|V_{kp}|^2$ in equation (7.35) which involves the two atom correlation function, described by the structure factor $S(K)$, in a quite fundamental way.

Obviously, we must now think of the potential scattering the electron under consideration from a state with wave vector $\mathbf{k}$ to a state $\mathbf{p}$ as the total potential created by all the ions (appropriately screened of course). We shall choose to express the total potential field as

$$V(\mathbf{r}) = \sum_{R_i} V_l(\mathbf{r} - \mathbf{R}_i) \qquad (7.36)$$

where we have in mind the localized potential of a single ion when we write V_l in (7.36).

We next form the matrix elements of $V(\mathbf{r})$ between the plane wave states $\Omega^{-\frac{1}{2}} e^{i\mathbf{k}\cdot\mathbf{r}}$ and $\Omega^{-\frac{1}{2}} e^{i\mathbf{p}\cdot\mathbf{r}}$ and we find

$$V_{kp} = \Omega^{-1} \int e^{-i\mathbf{k}\cdot\mathbf{r}} \sum_{R_i} V_l(\mathbf{r} - \mathbf{R}_i)\, e^{i\mathbf{p}\cdot\mathbf{r}}\, d\mathbf{r}$$

$$= \Omega^{-1} \sum_{R_i} \exp\left[i(\mathbf{k}-\mathbf{p})\cdot\mathbf{R}_i\right] \int \exp[i(\mathbf{p}-\mathbf{k})\cdot\mathbf{r}]\, V_l(\mathbf{r})d\mathbf{r}$$

$$= U_{k-p} N^{-1} \sum_i e^{i(\mathbf{k}-\mathbf{p})\cdot\mathbf{R}_i} \qquad (7.37)$$

where $U_{k-p} \equiv U_K$ is given by

$$U_K = N\Omega^{-1} \int e^{i\mathbf{K}\cdot\mathbf{r}} V_l(\mathbf{r})\, d\mathbf{r}. \qquad (7.38)$$

Liquid Metals

In the basic resistivity formula (7.35), we now have simply to substitute for V_{kp}, noting that

$$S(\mathbf{K}) = \frac{1}{N} \left| \sum_i \exp{(i\mathbf{K}\cdot\mathbf{R}_i)} \right|^2,\tag{7.39}$$

a slight variant on, but quite equivalent to, the results of Chapter 2, and we find by a simple change of variable in eqn. (7.35) the final result, in terms of v_f and the ion density ϱ_0,

$$\varrho_{\text{liquid}} = \frac{3\pi}{\hbar e^2 v_f^2 \varrho_0} \int_0^1 S(K) \, |U(K)|^2 \, 4\left(\frac{K}{2k_f}\right)^3 d\left(\frac{K}{2k_f}\right).\tag{7.40}$$

In principle, we can calculate the electrical properties of a liquid metal from this formula. We have seen that the structure factor $S(K)$ is known with some accuracy from experiment (see Figs. 2 and 3 as examples of results for Sn and Tl). Once we have made a choice of $U(K)$, ϱ_{liquid} follows.

7.6 Choice of scattering potential U(K)

The detailed choice of $U(K)$ hinges on pseudopotential theory. Apart from the factor $N\Omega^{-1}$ in eqn. (7.38), we might take as a first estimate of U_K the Fourier components of the screened potential around a point ion carrying a charge equal to the valency Z of the liquid metal. Then, if we assume for simplicity the screened Coulomb form (3.13), we find

$$U_{\mathbf{K}}^{\text{point ion}} = -\frac{N}{\Omega} \frac{4\pi Z e^2}{K^2 + q^2}\tag{7.41}$$

and since NZ is the total number of conduction electrons

$$U_{\mathbf{K}}^{\text{point ion}} = \frac{-4k_f^3 e^2}{3\pi(K^2 + q^2)}.\tag{7.42}$$

If we let $K \to 0$ we find

$$U_0^{\text{point ion}} = -\frac{k_f^2}{3} e^2 a_0 = -\frac{2}{3} E_f.\tag{7.43}$$

This is a central result of the theory and even when we transcend the point-ion approximation, as indeed we must, we have the value $-\frac{2}{3}E_f$ for $U(K)$ in the long wave limit $K \to 0$. This result,

66

however, is only rigorous in a linear theory and it is still too early to say how quantitatively accurate it will prove, especially in the polyvalent metals.

We can write eqn. (7.41) in the alternative form

$$U_{\mathbf{K}} = \frac{U_{\text{bare}}}{\varepsilon(K)} \tag{7.44}$$

where $\varepsilon(K)$ is given by eqn. (3.33), or, more accurately by eqn. (3.34), and

$$U_{\text{bare}}^{\text{point-ion}} = \frac{-4\pi NZe^2}{\Omega K^2}. \tag{7.45}$$

In the simplest argument, we have to take account of the fact that the conduction electrons, in real liquid metals with core electrons, do not see a bare ion potential of the form (7.45), that is pure Coulombic, but a much weaker potential. This circumstance arises as a consequence of orthogonalization of the conduction electron wave functions to the core states, and as we shall see shortly for Na, instead of U_{bare} being everywhere negative as in eqn. (7.45) for point ions, a node is introduced. This node, a consequence of a finite core, can have a profound influence on the electrical resistivity, when it comes around $2k_f$ as it appears to do for Na. The considerations below leave little doubt that we shall have to calculate the Fourier components of $U(K)$ with rather alarmingly high accuracy, if we are to make the theory fully quantitative. It is appropriate to remark at this point that $U(K)$ could also be used to approximate the pair potential, generalizing the point ion model of Chapter 3 (see, for example, Ziman, 1964). Since the difference between oscillatory and exponentially damped potentials leads to only small quantitative differences in $\mathbf{K}$ space (see Fig. 10), we can expect the nature of the pair potential to depend very sensitively also on $U(\mathbf{K})$. However, this statement should be qualified to the extent that if the form (3.34) is used for the dielectric constant screening the bare ion potential, then oscillations will appear, of wavelength π/k_f, in the long range tail of the potential. It will be interesting to see whether a quantitative relation can be developed between $U(K)$ and the direct correlation function (multiplied by $-k_B T$) in **K** space, which is, of course, an observable quantity (cf. Figs. 4 and 7).

However, it is worthwhile at this stage in the theory to consider specific methods used so far, to determine $U(K)$. We may cite three approaches:

(i) The original work of Ziman (1961), in which $U(K)$ was separated into two parts:

$$U(K) = U_1(K) + U_2(K). \qquad (7.46)$$

U_1, referred to as the plasma term, arises primarily from the dielectric screening by the conduction electrons, while U_2 is due to the detailed ion-core structure. Ziman treated $U_1(K)$ as a function tending to $-\frac{2}{3}E_f$ in the long-wave limit (cf. eqn. (7.43)) and becoming small as $K \to 2k_f$. Further, he approximated $U_2(K)$ in the region around the Fermi sphere diameter by a constant value, given by $U_2(K_1) = \frac{1}{2}E_g$, where K_1 is the smallest non-zero reciprocal lattice vector in the solid and E_g is the band gap at the nearest zone face. For small K, Ziman assumed that $U_2 \to 0$. Ziman stressed that such a separation was arbitrary, but that it had the useful consequence that for small K, $U_2(K) \sim 0$, whereas for K near $2k_f$, $U_1 \sim 0$. Moreover, from the density fluctuation theory (cf. Appendix 1), $S(K)$ in the long-wave limit is constant and proportional to kT, whereas in the region around $2k_f$, $S(K)$ is strongly varying.

Wiser (1966) has recently pointed out that, while Ziman's is an interesting start on the problem, in fact in general, even for Na, which one might think is the ideal case for the 'plasma' term because of the small band gap, the part of the resistivity coming from the structure term is larger than from the plasma term. Furthermore, the K dependence of U_2 around $2k_f$ is very important. Thus, perhaps such a separation has served its purpose and we should attack $U(K)$ as a single entity. This is done in the work of Greene and Kohn (1965), now to be discussed.

(ii) The analysis of Greene and Kohn is semi-empirical again, and is closely related to earlier work by Kohn and Vosko (1960) on impurities in metals. They essentially expressed $U(K)$ in terms of phase shifts, and attempted to determine the s and p phase shifts empirically. As Wiser (1966) has pointed out, their analysis is equivalent to replacing $U(K)$ by

$$U(K) = -\frac{4E_f}{3\pi} \sum_l (2l+1)\,\eta_l P_l \left(1 - \frac{K^2}{2k_f^2}\right), \qquad (7.47)$$

where η_l is the phase shift of the lth partial wave, and $P_l(x)$ is the Legendre polynomial of order l. Greene and Kohn now argued that the series in l converges fairly rapidly. They truncated the sum at the f waves ($l = 3$) and used a calculated value for η_3. Of the three remaining phase shifts, one is fixed by the Friedel sum rule (for monovalent metals)

$$\frac{2}{\pi} \sum_l (2l+1)\,\eta_l = 1; \quad \eta_l \equiv \eta_l(k_f) \qquad (7.48)$$

which is equivalent to saying again that $U(0) = -\frac{2}{3}E_f$. Thus the problem of finding $U(K)$ for all K is reduced to determining η_1 and η_2 (independent of K).

Substituting in the basic resistivity formula, one finds, after integrating over K, the form

$$\varrho = \sum_{l,\,l'} A_{ll'}(T)\,(2l+1)\,\eta_l(2l'+1)\eta_{l'}, \qquad (7.49)$$

where $A_{ll'}(T)$ contains the temperature dependence of ϱ. The method of Greene and Kohn was then to argue that at any temperature, eqn. (7.49) represents an ellipse in the space of η_0 and η_1. They then determined the phase shifts by looking for the common point of intersection of these ellipses at different temperatures. For solid N a, they found $\eta_0 = 0.524$, $\eta_1 = 0.258$, $\eta_2 = 0.036$, $\eta_3 = 0.015$.

However, in spite of the significant results obtained by Greene and Kohn, Wiser has pointed out that the method is beset by some difficulties. Thus, if we use eqn. (7.47) at $K = 2k_f$, we find

$$U(2k_f) = -\left(\frac{4E_f}{3\pi}\right) \sum_l (-1)^l\,(2l+1)\,\eta_l. \qquad (7.50)$$

Using the Friedel sum rule, and taking $\eta_3 = 0.015$, we have

$$U(2k_f) = -\left(\frac{8E_f}{3\pi}\right) (0.68 - 3\eta_1). \qquad (7.51)$$

Liquid Metals

Even the sign of $U(2k_f)$ is in doubt, when η_1 is around 0.2, as it is for Na.

(iii) The third method is to use direct pseudopotential calculations (Harrison, 1963; Heine and Abarenkov, 1964). Wiser shows that for Na, the estimated errors in Harrison's $U(K)$ can change ϱ by more than a factor of 3. Sundström (1965) has calculated ϱ with the model potential of Heine and Abarenkov, who determined the unscreened pseudopotential from spectroscopic data of

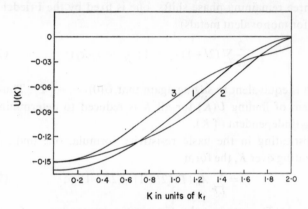

FIG. 21. Pseudopotentials for Na in Rydbergs
(1) From Heine–Abarenkov model potential
(2) From Harrison's potential
(3) From Greene–Kohn method

the free ions, in contrast to Harrison's direct calculation. The error in the Heine–Abarenkov potential is roughly the same as in Harrison's. The results for $U(K)$ for Na, obtained from methods (ii) and (iii), are shown in Fig. 21. The curves agree to about 0.03 Rydbergs at worst, but the calculated resistivities can differ by more than a factor of 2. We shall summarize below results obtained using the Heine–Abarenkov potentials for a variety of liquid metals (see Table 3), when we have discussed one further consequence of the Ziman theory.

TABLE 3

Values of resistivity and thermoelectric power

	$T°C$	ϱ in μ ohm cm		Q in μ volt/deg	
		Theory	Expt.	Theory	Expt.
Li	180	17	24	18	22
Na	100	9.4	9.7	$-$ 8	-10
K	65	32	13	-12	-16
Rb	40	14	22	$-$ 6	$-$ 7
Cs	30	13	37	$-$ 6	$+$ 6
Zn	460	44	37	$-$ 4	$-$ 6
Hg	23	77	98	$+$ 1	$-$ 5
Al	700	25	25	$-$ 4	$-$ 3
Tl	375	37	75	$-$ 4	
Pb	350	58	96	$-$ 5	$-$ 5
Bi	300	109	129	$-$ 4	$-$ 1

7.7 Thermoelectric power

In addition to the resistivity, it is worthwhile to consider briefly the results of the theory for the thermoelectric power Q at this point. We adopt the conventional representation of Q through a dimensionless quantity ξ defined by

$$\xi = Q \Big/ \left(\frac{\pi^2 k_B^2 T}{3eE_f} \right). \tag{7.52}$$

Then it can be shown (see, for example, Ziman, 1960) that if the conductivity σ is known as a function of energy in the region near to the Fermi level,

$$\xi = E_f \left[\frac{\partial \ln \sigma(E)}{\partial E} \right]_{E=E_f}. \tag{7.53}$$

Using the theory developed in detail above, it is a straightforward matter to find ξ in terms of the Fourier transform of the scattering potential $U(K)$ and the structure factor $S(K)$. The result takes the form (for a local pseudopotential)

$$\xi \doteq 3 - \frac{2S(2k_f)\,|\,U(2k_f)\,|^2}{\langle SU^2 \rangle} \tag{7.54}$$

Liquid Metals

where

$$\langle SU^2 \rangle = 4 \int_0^1 S(K) \, |U(K)|^2 \left(\frac{K}{2k_f}\right)^3 d\left(\frac{K}{2k_f}\right). \qquad (7.55)$$

This latter quantity, from equation (7.40), essentially determines the liquid resistivity ϱ and relates ϱ to the thermopower, the measured structure factor at $2k_f$ and $U(2k_f)$. The most detailed calculations available to date are again those of Sundström (1965) using the model potentials of Heine and Abarenkov. We give a selection of her results in Table 3 along with the experimental values. Considering the uncertainties in the basic quantities $S(K)$ and $U(K)$, the agreement seems as good as could reasonably be expected.

7.8 Hall coefficient

We remark briefly that the Hall coefficient in liquid metals has been found to be well described by the usual formula

$$R = \frac{1}{ne} \qquad (7.56)$$

where n comes out to agree with the value obtained from the valence (e.g. four conduction electrons per atom for Sn). While the result (7.56) is true for free electrons, it follows under much less restrictive assumptions. Thus, a parabolic relation between energy E and wave number k is not required, but only that the dispersion law $E(k)$ and the relaxation time $\tau(k)$ are independent of the direction of $\mathbf{k}$. These assumptions seem well borne out in many liquid metals.

Attempts have been made to work out an improved theory of the Hall effect in liquid metals (Kubo, 1964; Springer, 1964) but the corrections due to the disorder scattering seem small, when viewed in the light of the state of knowledge in other aspects of the problem. However, pseudopotential methods do not look very promising as a means of calculating such corrections to eqn. (7.56).

7.9 Blurring of Fermi surface

The disorder scattering referred to above is bound to have some effect on the Fermi surface, and will tend to blur it. But all the arguments so far presented have been based on the assumption that it is still sharply defined.

That this is indeed so, to a good approximation, can be seen rather directly from the conductivity results we have been discussing. Calculating the electronic mean free path directly from the measured conductivity, we can then argue that this is the range over which an electronic wave packet loses its coherence. Thus the spread in k values for electrons taking part in the conduction process (i.e. around k_f) is given by $\Delta k \sim \dfrac{1}{\lambda}$. $\Delta k/k_f$ can then be crudely estimated, as shown in Table 4. The blurring is never more than 10 per cent or so, and indeed is very much less for Na.

TABLE 4

Blurring of Fermi surface in liquid metals

	$\lambda(\text{Å})$	$\Delta k(\text{Å}^{-1})$	$k_f(\text{Å}^{-1})$	$\Delta k/k_f$
Na	157	0.0064	0.89	0.007
Hg	7	0.14	1.34	0.10
Pb	6	0.17	1.54	0.11

But the evidence which could, in principle, decide the question we are asking about the existence of a sharp edge at the Fermi surface most directly, is that available from the positron annihilation experiments. In particular, Gustafson, Mackintosh and Zaffarano (1963) have reported recently on positron annihilation measurements in both solid and liquid Hg, and their results have been interpreted in terms of an electronic momentum distribution $P(k)$, which, in the liquid, is considerably more diffuse than in the solid. Adopting a form of $P(k)$ suggested by analogy with thermal exci-

Liquid Metals

tation, namely

$$P(k) = \frac{A}{\exp\left(\dfrac{k^2 - 1}{\Delta}\right) + 1} \qquad (7.57)$$

with k in units of k_f, they find $\Delta \approx 0.2$.

While this diffuseness might have been supposed to arise from the disorder scattering of the electrons, there are two points which can be made in connection with it:

(i) The diffuseness may be due to the thermalization process, or the effect of disorder on the positron wave function.

(ii) While the diffuseness is roughly compatible with the mean free path given in Table 4, Mott (1966) has argued that the mean free path in liquid Hg is longer than that obtained from the elementary conductivity argument used above.

If the diffuseness of $P(k)$ in Hg were indeed as large as that supposed by Gustafson *et al.*, then the considerations above would be in jeopardy. For Gaskell and March (1963) have shown that such a blurring of around 20 per cent would damp out the oscillatory effects in the ion–ion interaction. The results of their calculations of the screening round a point ion in a Fermi gas of electron density appropriate to liquid Hg are displayed in Fig. 22. Curve 1 gives the result derived from (3.37) and (3.34) by Fourier transform, curve 2 shows the damped oscillatory potential when the Fermi surface is blurred by 10 per cent, while curve 3 indicates that the oscillations are effectively suppressed if the blurring is as large as 20 per cent. The theory is admittedly approximate but it shows us that oscillatory interactions are very sensitively dependent on the sharpness of the Fermi surface. It should be stressed that a blurring of around 10 per cent would present no difficulty of this kind, as can be seen by examination of Fig. 22.

If the Fermi surface is not sharp, then we suggest that the stability of metallic liquid Hg itself presents some difficulty and we therefore must assume that the positron annihilation results in the liquid require some degree of reinterpretation (see, however, Faber, 1966).

74

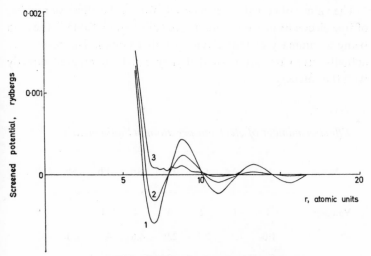

FIG. 22. Screened ion potential in liquid Hg
(1) Sharp Fermi surface
(2) Blurring of 10 per cent
(3) Blurring of 20 per cent

7.10 Optical properties

While our main object of outlining the present theory of the d.c. conductivity of liquid metals has been achieved, we shall briefly comment here on the optical properties of liquid metals (for a recent review, see Faber, 1966).

Experimentally the optical properties of liquid metals are easier to measure than those of solids. Also, with few exceptions known as yet, the a.c. conductivity is adequately represented by the Drude formula

$$\sigma(\omega) = \frac{n^* e^2 \tau}{m(1 + \omega^2 \tau^2)} \qquad (7.58)$$

and $\omega\tau \sim 1$ in the visible, contrasted with the case of solids where $\omega\tau \sim 1$ lies in the far infrared. Equation (7.58), of course, is the real part of the complex conductivity, the imaginary part being derivable from it, in principle, using the Kramers–Kronig relations.

Liquid Metals

The values obtained experimentally for n^*, the effective number of free electrons per unit volume are collected in Table 5, the data being a summary of that given by Faber (1966). The results are actually given per atom so that they can be compared directly with the valency.

TABLE 5

Effective number of electrons per atom in liquid metals

Metal	K	Ag	Cd	Ga	Sn	Pb	Sb	Bi
Valency	1	1	2	3	4	4	5	5
n^*	1.0	1.1	2.1	2.9	4.6	4.7	6.1	5.3

The data available on the optical properties of liquid metals is then seen to agree well with free-electron theory (a notable exception, however, is liquid Hg, which does not obey the Drude formula (7.58)). This, at first sight, is surprising because there is some evidence that the density of states in liquid metals can often be non-free-electron like. However, the theory of electrons in disordered structures as given by Edwards (1962) shows that the optical properties should not be sensitive to the density of states. Essentially, what Edwards showed was that the conductivity of a liquid metal can still be calculated using the free electron model, even when the density of states departs substantially from the free-electron value (again, liquid Hg seems exceptional, as discussed by Mott, 1966). This is the underlying reason why we can still use the Ziman theory outlined above, and summarized in equation (7.40), with v_f calculated as in free-electron theory.

7.11 Alloys

We turn finally to consider briefly how the theory discussed above can allow an understanding of the general features of binary metallic alloys. To solve the problem completely, even within the approximate framework used in this chapter, we should require, for $A-B$ alloys, two pseudopotentials and three partial structure factors S_{AA}, S_{BB} and S_{AB}. Only recently, in an important experiment by Enderby, North and Egelstaff (1966) on Cu–Sn alloys, have such structure factors become available.

Nevertheless, the work of Faber and Ziman (1965) appears to lead to a general understanding of some hitherto rather puzzling features of alloy resistivities. It is well known that in the solid state, the resistivity of a binary alloy obeys two rules. The first, due to Nordheim (1931), is that the dependence of resistivity ϱ on atomic concentration c should be roughly like $c(1-c)$. Thus, a plot of ϱ against c should be convex upwards, and this is often a very pronounced effect. The second rule, due to Linde (1932), is that $d\varrho/dc$, for dilute alloys, should be roughly proportional to the valence difference between solute and solvent.

When the behaviour of a variety of liquid alloys is studied, it soon becomes apparent that these rules are not obeyed. It is only when the solvent is a monovalent metal such as Cu that the resistivity of the liquid alloy behaves in the same general way as the solid.

To expose the main features of the argument of Faber and Ziman (1965), we recall that the resistivity in Born approximation may be written

$$\varrho = \frac{3\pi}{\hbar e^2} \frac{1}{\varrho_0 v_f^2} \langle SU^2 \rangle \qquad (7.59)$$

where the average is immediately defined by comparison with equation (7.55). When we deal with a binary alloy, as we remarked above, we need to know the $A-A$, $B-B$ and $A-B$ structure factors. However, if we assume that the two constituents have the same atomic volume and valency, then plausibly $S_{AA} = S_{BB} = S_{AB} = S$ say, then we can proceed to develop a rough theory.

If we assign localized pseudopotentials U_A and U_B to the atoms A and B, then we may write

$$\varrho = \varrho_1 + \varrho_2 \qquad (7.60)$$

where

$$\varrho_1 = [(3\pi/\hbar e^2)/(\varrho_0 v_f^2)] \langle (1-c)SU_A^2 + cSU_B^2 \rangle \qquad (7.61)$$

77

Liquid Metals

and

$$\varrho_2 = [(3\pi/\hbar e^2)/(\varrho_0 v_f^2)] \langle c(1-c)(1-S)(U_A - U_B)^2 \rangle. \quad (7.62)$$

The terms quadratic in the concentration c are now seen to appear only in ϱ_2: these are the terms responsible for Nordheim's rule. In contrast, ϱ_1 should vary in a roughly linear fashion between ϱ_A and ϱ_B, the resistivities of the pure solvent and pure solute respectively.

The overall behaviour to be found in liquid alloys then depends on whether ϱ_1 or ϱ_2 is dominant. This in turn depends largely on the magnitude of the structure factor S.

A rather striking confirmation of these ideas is contained in the experiments of Enderby and Howe (private communication) on a Ag–Au liquid alloy. These workers have measured the thermopower over the entire range of concentration. In this system, the assumptions discussed above are most probably validated and equations (7.60)–(7.62) can be used to calculate the thermopower via equation (7.53), across the entire phase diagram.

The results fit the experiments in a fully quantitative way. We can expect a good deal more progress in this important area now that partial structure factors are accessible from diffraction experiments.

CHAPTER 8

Liquid dynamics

IN CHAPTER 1, we considered the way in which the radial distribution function $g(r)$, or the Fourier transform of $g(r)-1$, the structure factor $S(K)$, is related to X-ray scattering.

In the more advanced account of neutron scattering from liquids taken up in the present chapter, we shall see that neutrons can give us additional basic information about the dynamics of atoms in liquids. The theory so far has developed along two main lines:

(i) Formal properties of the generalized correlation functions, introduced into the theory by Glauber (1955) and van Hove (1954).

(ii) Simple models to represent the measured data on liquids. So far lacking is a more basic discussion of the way in which we can calculate the correlation function from a knowledge of the interatomic or interionic forces. This is not surprising, for, as we shall see below, the van Hove correlation function is a generalization of the structure factor $S(K)$. Even $S(K)$ itself, for realistic forces, has been obtained only by a virial expansion or from the approximate theories of Chapter 4.

8.1 Definition of van Hove correlation function

We will introduce the van Hove correlation function in an intuitive way, and will later point out its intimate connection with neutron scattering. We argue purely classically at first. Suppose that we define $G(\mathbf{r}, t)$ as the average density of atoms at the point $\mathbf{r}$ at time t, if an atom was at the origin $\mathbf{r} = 0$ at time $t = 0$.

Liquid Metals

Thus, it gives us the correlation in the positions of two atoms, which may or may not be different, at different times. This function may then be written in the form

$$G(\mathbf{r}, t) = \frac{1}{N} \left\langle \sum_{i,\,j=1}^{N} \delta[\mathbf{r}+\mathbf{R}_i(0)-\mathbf{R}_j(t)] \right\rangle. \tag{8.1}$$

The average of the δ function involved in (8.1) is obviously the probability that at time t the jth atom will be at $\mathbf{r}$ with respect to the position of the ith atom at time $t = 0$. We then sum this probability over j and average over i. $G(\mathbf{r}, t)$ is the space–time pair correlation function.

The quantum mechanical generalization of (8.1) is in fact

$$G_{\text{quantum}}(\mathbf{r}, t) = \frac{1}{N} \left\langle \sum_{i,\,j=1}^{N} \int d\mathbf{r}' \, \delta[\mathbf{r}+\mathbf{R}_i(0)-\mathbf{r}'] \, \delta[\mathbf{r}'-\mathbf{R}_j(t)] \right\rangle$$
$$\tag{8.2}$$

where $\mathbf{R}_i(0)$ and $\mathbf{R}_j(t)$ are non-commuting Heisenberg operators. If this failure to commute is ignored, it proves possible to integrate over $\mathbf{r}'$, and then (8.1) is regained.[†]

An alternative and sometimes useful form for $G(\mathbf{r}, t)$ is obtained by introducing the density operator $\varrho(\mathbf{r}, t)$ of the atoms at the point $\mathbf{r}$ at the time t:

$$\varrho(\mathbf{r}, t) = \sum_{i=1}^{N} \delta[\mathbf{r}-\mathbf{R}_i(t)]. \tag{8.3}$$

If we employ this form in the definition (8.2), and change the origin by substituting $\mathbf{r}'' = \mathbf{r}' - \mathbf{r}$, then we obtain

$$G_{\text{quantum}}(\mathbf{r}, t) = \frac{1}{N} \left\langle \int d\mathbf{r}'' \, \varrho(\mathbf{r}'', 0) \, \varrho(\mathbf{r}''+\mathbf{r}, t) \right\rangle. \tag{8.4}$$

Thus, we can interpret $G(\mathbf{r}, t)$ as the space–time correlation of the density ϱ.

If we now take explicit account of the homogeneity of the liquid, then the integrand in (8.4) is independent of $\mathbf{r}''$ and we find

$$G(\mathbf{r}-\mathbf{r}', t-t') = \frac{1}{\varrho_0} \langle \varrho(\mathbf{r}', t') \, \varrho(\mathbf{r}, t) \rangle. \tag{8.5}$$

[†] $G_{\text{quantum}}(\mathbf{r}, t)$ is in general complex and has no immediate probabilistic interpretation.

Now we follow van Hove, and take the diagonal terms $i = j$ out of the sum over i and j in (8.2), when we obtain

$$G(\mathbf{r}, t) = G_s(\mathbf{r}, t) + G_d(\mathbf{r}, t) \qquad (8.6)$$

where

$$G_s(\mathbf{r}, t) = \frac{1}{N} \left\langle \sum_{i=1}^{N} \int d\mathbf{r}' \, \delta[\mathbf{r} + \mathbf{R}_i(0) - \mathbf{r}'] \, \delta[\mathbf{r}' - \mathbf{R}_i(t)] \right\rangle \quad (8.7)$$

and

$$G_d(\mathbf{r}, t) = \frac{1}{N} \left\langle \sum_{i \neq j=1}^{N} \int d\mathbf{r}' \, \delta[\mathbf{r} + \mathbf{R}_i(0) - \mathbf{r}'] \, \delta[\mathbf{r}' - \mathbf{R}_j(t)] \right\rangle. \quad (8.8)$$

By this separation, we can interpret G_s as the correlation function which tells us the probability that a particle which was at the origin at time $t = 0$, will be at $\mathbf{r}$ at time t. The part G_d obviously refers to the analogous conditional probability of finding a different atom at $\mathbf{r}$ at time t.

Let us now investigate the correlation functions at time $t = 0$. Going back to (8.2) and noting that $\mathbf{R}_i(0)$ and $\mathbf{R}_j(0)$ commute, we can integrate over $\mathbf{r}''$ (cf. remarks after (8.2)) and we find

$$G_{\text{quantum}}(\mathbf{r}, 0) = \frac{1}{N} \left\langle \sum_{i, j=1}^{N} \delta[\mathbf{r} + \mathbf{R}_i(0) - \mathbf{R}_j(0)] \right\rangle. \quad (8.9)$$

Splitting this according to (8.6) we find almost immediately

$$G_s(\mathbf{r}, 0) = \delta(\mathbf{r}) \qquad (8.10)$$

$$G(\mathbf{r}, 0) = \delta(\mathbf{r}) + \varrho_0 g(\mathbf{r}) \qquad (8.11)$$

where $g(\mathbf{r})$ is given by

$$\varrho^0 g(\mathbf{r}) = \frac{1}{N} \sum_{i \neq j=1}^{N} \langle \delta(\mathbf{r} + \mathbf{R}_i - \mathbf{R}_j) \rangle, \qquad (8.12)$$

the radial distribution function, of which we have made widespread use earlier in this volume.

In the limit of long times, we can assume there is no correlation between positions of particles. Thus, in (8.2), we can replace the average of the product of the δ-functions by the product of the averages

$$\sum_{i, j=1}^{N} \langle \delta[\mathbf{r} + \mathbf{R}_i(0) - \mathbf{r}'] \, \delta[\mathbf{r}' - \mathbf{R}_j(t)] \rangle$$

$$\approx \left\langle \sum_{i=1}^{N} \delta[\mathbf{r} + \mathbf{R}_i(0) - \mathbf{r}'] \right\rangle \left\langle \sum_{j=1}^{N} \delta[\mathbf{r}' - \mathbf{R}_j(t)] \right\rangle. \quad (8.13)$$

81

Liquid Metals

Thus, for large **r** or large t we may write

$$G_{\text{quantum}}(\mathbf{r}, t) = \frac{1}{N} \int d\mathbf{r}' \, \varrho(\mathbf{r}' - \mathbf{r}) \, \varrho(\mathbf{r}'), \qquad (8.14)$$

which is simply the density–density correlation function. For systems with long-range order, $\varrho(\mathbf{r})$ and $G(\mathbf{r}, t)$ are periodic in space. For a fluid $\varrho_0 = \dfrac{N}{\Omega}$ where Ω is the volume, and therefore

$$G_{\text{quantum}}(\mathbf{r}, \infty) \approx \varrho_0. \qquad (8.15)$$

Similarly for the self-correlation function we can show that for a homogeneous system

$$G_s(\mathbf{r}, \infty) = \frac{1}{\Omega}, \qquad (8.16)$$

which tends to zero as Ω tends to infinity. This is in marked contrast to the situation in which atoms are not free to move far from

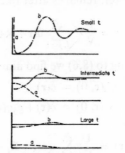

FIG. 23. Schematic form of 'self' and 'distinct' parts of van Hove correlation function

(a) $G_s(\mathbf{r}, t)$
(b) $G_d(\mathbf{r}, t)$

some 'lattice' sites. In this case, appropriate to solids when we neglect diffusion, $G_s(\mathbf{r}, \infty) \neq 0$.

For short times $G_s(\mathbf{r}, t)$ approximates to a δ function according to (8.10), while $G_d(\mathbf{r}, t)$ is approximately the pair correlation function $g(r)$. As $t \to \infty$, $G_s(\mathbf{r}, t) \to 0$ and $G_d(\mathbf{r}, t) \to \varrho_0$, and these forms are shown schematically in Fig. 23.

8.2 Models of self-correlation function

We now take some special cases.

8.2.1 FREE PARTICLE

Classically, we have that

$$\mathbf{r} = \mathbf{v}t \tag{8.17}$$

and thus

$$G_s(\mathbf{r}, t) = \delta(\mathbf{r} - \mathbf{v}t)$$

$$= \delta\left(\mathbf{r} - \frac{\mathbf{p}t}{M}\right), \tag{8.18}$$

where $\mathbf{v}$ is the velocity, $\mathbf{p}$ the momentum and M the atomic mass Due to quantum effects, this is not correct. The self-term $G_s(\mathbf{r}, t)$. has a Gaussian shape, though its maximum remains at $\mathbf{r} = \mathbf{v}t$. Explicitly

$$G_s(\mathbf{r}, t) = \left(2\pi \frac{ht}{iM}\right)^{-\frac{3}{2}} \exp\left\{\frac{-\left(\mathbf{r} - \frac{\mathbf{p}}{M}t\right)^2}{2\left(\frac{ht}{iM}\right)}\right\}. \tag{8.19}$$

which reduces to (8.18) as $h \to 0$.

8.2.2 PERFECT GAS

For a perfect gas at temperature T we have the result

$$G_s(\mathbf{r}, t) = [2\pi\Gamma(t)]^{-\frac{3}{2}} \exp\left\{\frac{-r^2}{2\Gamma(t)}\right\}, \tag{8.20}$$

where

$$\Gamma(t) = \frac{k_B T}{M} t \left(t - \frac{i\hbar}{k_B T}\right). \tag{8.21}$$

For $t \ll \frac{\hbar}{k_B T}$, $\Gamma(t) \approx -\frac{i\hbar}{M} t$ and we have the problem dominated by quantum mechanics, while for $t \gg \frac{\hbar}{k_B T}$, $\Gamma(t) = \frac{k_B T t^2}{M} = \langle (v_x t)^2 \rangle$, which is purely classical.

8.2.3 DIFFUSION

In this model, we suppose that if an atom of the liquid was situated at the origin at time $t = 0$, the probability of finding it at r after a long time can be obtained by solving the diffusion equation, as was first proposed by Vineyard (1958). In terms of the selfdiffusion coefficient D we find

$$G_s(\mathbf{r}, t) = \frac{1}{(4\pi D |t|)^{\frac{3}{2}}} \exp \left(\frac{-r^2}{4D |t|} \right). \qquad (8.22)$$

8.2.4 INTERPOLATION MODEL

Forms (8.20) and (8.22) have considerable similarity, and suggest that $\Gamma(t)$, the mean square atomic displacement, may be sketched roughly as a function of time using our models as a basis.

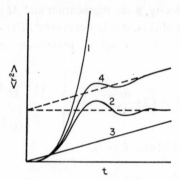

FIG. 24. Schematic form of mean square atomic displacement as a function of time
(1) Perfect gas
(2) Crystal
(3) Diffusion model
(4) Liquid

Thus we show in Fig. 24, $\Gamma(t)$ against t for the models (ii) and (iii) (curves 1 and 3 respectively). For a crystal, $\Gamma(t)$ must remain finite as $t \to \infty$ (curve 2) while curve 4 is a rough sketch of what the result may be like in a liquid. We shall be discussing the quantitative form of the mean square displacement for insulators and for metals later in the chapter.

8.3 Neutron scattering law

The neutron scattering cross-section per particle per unit solid angle and per unit of outgoing energy is (see, for example, Kittel, 1963; March, Young and Sampanthar, 1967)

$$\frac{d^2\sigma}{d\Omega\, d\omega} = \frac{\hbar k}{Nk_0} \sum_i p_i \sum_f \left| \left\langle f \left| \sum_n a_n e^{i\mathbf{K}\cdot\mathbf{r}_n} \right| i \right\rangle \right|^2 \delta(E_f - E_i - \hbar\omega).$$

(8.23)

Here i and f denote the initial and final states of the system with energy E_i and E_f, $\mathbf{k}_0$ and $\mathbf{k}$ the wave vectors of the incoming and outgoing neutron respectively. N is the total number of atoms, a_n is the scattering length of the nth atom and p_i is the relative population of the initial state i.

Further
$$\hbar\omega = \frac{\hbar^2}{2m}(k_0^2 - k^2)$$
(8.24)

and
$$\hbar\mathbf{K} = \hbar(\mathbf{k}_0 - \mathbf{k})$$
(8.25)

are respectively the energy and momentum transfer from the neutron to the system, where m is the neutron mass.

It may then be shown that (8.23) can be transformed into

$$\frac{d^2\sigma}{d\Omega\, d\omega} = \frac{k}{k_0}\{a_{\text{inc}}^2 S_{\text{inc}}(\mathbf{K}, \omega) + a_{\text{coh}}^2 S_{\text{coh}}(\mathbf{K}, \omega)\}$$
(8.26)

where $a_{\text{inc}}^2 = \langle a^2\rangle - \langle a\rangle^2$, $a_{\text{coh}} = \langle a\rangle$ and the incoherent and coherent scattering functions S can be expressed as Fourier transforms of the so-called intermediate scattering functions F, which are, in fact, time dependent correlation functions. The relations between S and F are then as follows:

$$S_{\text{inc}}(\mathbf{K}, \omega) = \frac{1}{2\pi}\int_{-\infty}^{\infty} e^{-i\omega t} F_s(\mathbf{K}, t)\, dt$$

$$= \frac{1}{2\pi}\int_{-\infty}^{\infty} e^{-i\omega t} \left\langle e^{-i\mathbf{K}\cdot\mathbf{r}_1(0)} e^{i\mathbf{K}\cdot\mathbf{r}_1(t)} \right\rangle dt.$$
(8.27)

$$S_{\text{coh}}(\mathbf{K}, \omega) = \frac{1}{2\pi}\int_{-\infty}^{\infty} e^{-i\omega t} F(\mathbf{K}, t)\, dt$$

$$= \frac{1}{2\pi}\int_{-\infty}^{\infty} e^{-i\omega t} \sum_n \left\langle e^{-i\mathbf{K}\cdot\mathbf{r}_1(0)} e^{i\mathbf{K}\cdot\mathbf{r}_n(t)} \right\rangle dt.$$
(8.28)

Liquid Metals

As in (8.2), the coordinate $r_n(t)$ is to be interpreted as the corresponding Heisenberg operator (cf. Messiah, 1961) and the thermal average of an operator Q is given by

$$\langle Q \rangle = \sum_i p_i \langle i | Q | i \rangle. \tag{8.29}$$

Having defined the correlation functions S and F, it only remains to make contact with the coordinate space forms of section 8.1. The results may be written in the form

$$S_{\text{inc}}(\mathbf{K}, \omega) = \frac{1}{2\pi} \int\int dt \, d\mathbf{r} \, e^{i(\mathbf{K} \cdot \mathbf{r} - \omega t)} G_s(\mathbf{r}, t) \tag{8.30}$$

and

$$S_{\text{coh}}(\mathbf{K}, \omega) = \frac{1}{2\pi} \int\int dt \, d\mathbf{r} \, e^{i(\mathbf{K} \cdot \mathbf{r} - \omega t)} G(\mathbf{r}, t). \tag{8.31}$$

8.4 Small time expansion of intermediate scattering function

From the definition of $F_s(\mathbf{K}, t)$ given in (8.27) we have

$$F_s(\mathbf{K}, t) = \langle e^{i\mathbf{K} \cdot [\mathbf{r}_1(t) - \mathbf{r}_1(0)]} \rangle, \tag{8.32}$$

where we have assumed that we can restrict ourselves to the classical case. This is usually adequate in simple liquids, but we can always estimate the quantum corrections later. The classical assumption amounts to neglecting the commutator of the operators $\mathbf{r}_1(0)$ and $\mathbf{r}_1(t)$ in (8.27), and instead of employing (8.29) as the thermal average, we simply average over a classical canonical ensemble in the phase space.

If the classical Hamiltonian has the form

$$H = \frac{1}{2M} \sum_i \mathbf{p}_i^2 + \Phi(\mathbf{r}_1, \ldots, \mathbf{r}_N), \tag{8.33}$$

then we may write

$$F_s(\mathbf{K} \, t) = \frac{\int e^{-\beta H} e^{i\mathbf{K} \cdot [\mathbf{r}_1(t) - \mathbf{r}_1(0)]} \, d\tau}{\int e^{-\beta H} \, d\tau}; \quad \beta = 1/k_B T. \tag{8.34}$$

The integrations in (8.34) are over the whole phase space for N atoms.

86

Obviously we can re-express (8.34) in the form

$$F_s(\mathbf{K}, t) = \left\langle e^{iK\int_0^t \dot{x}_1(t_1)\,dt_1} \right\rangle \tag{8.35}$$

where we have taken the x axis along the direction of $\mathbf{K}$.

We now expand (8.35) in the form

$$F_s(\mathbf{K}, t) = \exp\{-K^2\gamma_1(t) + K^4\gamma_2(t) + \ldots\} \tag{8.36}$$

and we then find, after some manipulation,

$$\gamma_1(t) = \int_0^t (t - t_1)\langle \dot{x}_1(0)\,\dot{x}_1(t_1)\rangle\,dt_1$$

$$= \tfrac{1}{6}\langle[\mathbf{r}_1(t) - \mathbf{r}_1(0)]^2\rangle. \tag{8.37}$$

Thus, from (8.37) we see that $\gamma_1(t)$ is directly related to the mean square displacement of a particle.

Since it can be shown (cf. Schofield, 1961) that $\gamma_2(t)$ in (8.36) is $\propto t^8$ for small t, it seems a reasonable first approximation, for times that are not too large, to write

$$F_s(\mathbf{K}, t) \doteq \exp\{-K^2\gamma_1(t)\}. \tag{8.38}$$

Focusing attention therefore on the calculation of $\gamma_1(t)$, to use, for example, in the approximate form (8.38), we notice from (8.37) that

$$\ddot{\gamma}_1(t) = \langle \dot{x}_1(0)\,\dot{x}_1(t)\rangle. \tag{8.39}$$

8.5 Expansion of velocity correlation function

It is now a straightforward matter to expand the velocity correlation function in (8.39) for small times t. By simple integration we then find

$$\gamma_1(t) = \frac{t^2}{2\beta M} - \langle \ddot{x}_1(0)^2\rangle\frac{t^4}{4!} + \langle \dddot{x}_1(0)^2\rangle\frac{t^6}{6!} - \ldots \tag{8.40}$$

It is also well known that for large times

$$\gamma_1(t) \to Dt + \text{constant}, \tag{8.41}$$

where D is the self-diffusion coefficient (cf. eqn. (8.22)).

87

Liquid Metals

Explicitly, if we write the total potential energy $\Phi(\mathbf{r}_1, \ldots, \mathbf{r}_N)$ in terms of pair potentials

$$\Phi(\mathbf{r}_1, \ldots, \mathbf{r}_N) = \sum_{i<k} \phi(r_{ik}) \tag{8.42}$$

then the second term in (8.40) becomes

$$\langle \ddot{x}_1(0)^2 \rangle = \frac{4\pi\varrho_0}{3\beta M^2} \int_0^\infty g(r) \left[\phi''(r) + \frac{2}{r}\,\phi'(r) \right] r^2\,dr. \tag{8.43}$$

The next term in (8.40) may also be written down, but it involves the three-atom correlation function and we refer the reader to the work of Nijboer and Rahman (1966) for further details.

An alternative form of $F_s(\mathbf{K}, t)$ which we can transform to obtain $S_{\text{inc}}(\mathbf{K}, \omega)$ is given from (8.38) and (8.40) as

$$F_s(\mathbf{K}, t) = e^{\frac{-K^2 t^2}{2\beta M}} \left[1 + K^2 \langle \ddot{x}_1(0)^2 \rangle \frac{t^4}{4!} - K^2 \langle \dddot{x}_1(0)^2 \rangle \frac{t^6}{6!} + \ldots \right]. \tag{8.44}$$

8.6 Large K expansion of self-correlation function

From (8.27) and (8.44) we can now calculate $S_{\text{inc}}(\mathbf{K}, \omega)$. Introducing the quantity $\alpha = (K^2/2\beta M)$ to simplify the notation, we find

$$S_{\text{inc}}(K, \omega) = (4\pi\alpha)^{-\frac{1}{2}} e^{-\frac{\omega^2}{4\alpha}} \left\{ 1 + \frac{\beta^2 M^2}{4!\,K^2} \langle \ddot{x}_1(0)^2 \rangle H_4 \left(\frac{\omega}{\sqrt{2\alpha}} \right) \right.$$
$$\left. + \frac{\beta^3 M^3}{6!\,K^4} \langle \dddot{x}_1(0)^2 \rangle H_6 \left(\frac{\omega}{\sqrt{2\alpha}} \right) + \ldots \right\} \tag{8.45}$$

where the Hermite polynomials H_n are defined by

$$H_n(x) = (-1)^n e^{\frac{x^2}{2}} \frac{d^n}{dx^n} \left(e^{-\frac{1}{2}x^2} \right). \tag{8.46}$$

A discussion similar to that described above was given by Nelkin and Parks (1960). It will be seen that (8.45) is a large K expansion of $S_{\text{inc}}(K, \omega)$.

8.7 Comparison with machine calculations and with experiment

Only a rather limited comparison of the theory presented above can be made with experiment. However, using the method of molecular dynamics, Rahman (1964) has been able to calculate $\gamma_1(t)$, or the mean square displacement, by machine. In this work, a few hundred atoms only are considered, with a Lennard-Jones type of interaction. Essentially then, the classical Newtonian equations are solved, with periodic boundary conditions.

Although the Lennard-Jones force law is not appropriate to liquid metals, we shall briefly discuss Rahman's results for this case, with parameters appropriate to liquid argon.

8.7.1 MEAN SQUARE DISPLACEMENT IN ARGON

We show in Fig. 25 the results obtained by Rahman for the mean square atomic displacement in A at 85.5°K. Similarly, the velocity correlation function in (8.39) can be obtained from the machine calculations, along with the small time expansions. The results are displayed in Fig. 26. It will be seen that little is gained by adding the t^4 term in the expansion of (8.39); indeed it seems as if the series is more useful (presumably it is an asymptotic expansion) when truncated at the t^2 term. Thus the velocity correlation

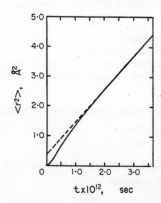

Fig. 25. Mean square displacement in liquid argon at 85.5°K
Asymptotic form is $6Dt + C$, with $D = 1.88 \times 10^{-5}$ cm^2 sec^{-1}
and $C = 0.36$Å^2

function cannot be approximated, even out to its first zero, by a small time expansion and further machine calculations for other liquids seem unavoidable.

The slope of $\langle r^2 \rangle$ at large times is related to the diffusion constant D through eqn. (8.41). This quantity appears to be somewhat sensitive to the nature of the force law, and we shall therefore turn now to consider the results of Paskin and Rahman (1966),

$$\frac{\langle \dot{x}_1(0)\, \dot{x}_1(t) \rangle}{\langle \dot{x}_1(0)\, \dot{x}_1(0) \rangle}$$

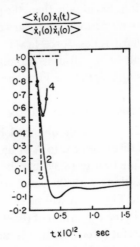

FIG. 26. Velocity correlation function $\dfrac{\langle \dot{x}_1(0)\, \dot{x}_1(t) \rangle}{\langle \{\dot{x}_1(0)\}^2 \rangle}$ for liquid argon at 85.5°K

(1) Perfect gas
(2) Machine calculation
(3) Small time expansion including t^2 term only
(4) Small time expansion including t^2 and t^4 term

who calculated D for some oscillatory potentials believed to be appropriate for liquid Na.

8.7.2 TRUNCATED OSCILLATORY POTENTIALS AND DIFFUSION COEFFICIENT IN SODIUM

It is obvious from the discussions of Chapter 5 that an acceptable pair potential $\phi(r)$ must generate the measured structure factor $S(K)$, and, in particular, account for the striking short-range nature of $\hat{f}(K)$, the Fourier transform of the direct correlation function in liquid metals.

For Na, however, we saw in Chapter 5 that the potential which generates the X-ray measurements and that which generates the neutron results are very substantially different.

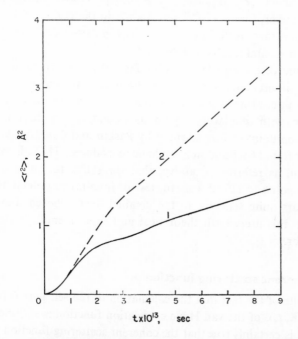

FIG. 27. Mean square displacement $\langle r^2 \rangle$ against time for liquid Na
(1) Paskin–Rahman potential (1). (Curve 1 of Fig. 15)
(2) Paskin–Rahman potential (2). (Curve 2 of Fig. 15)

A further criterion therefore which an acceptable pair potential must satisfy is that it should generate the self-diffusion constant. Using the molecular dynamics approach referred to above, together with the truncated oscillatory potentials for Na shown in curves 1 and 3 of Fig. 15, Paskin and Rahman (1966) have calculated the mean square displacement $\langle r^2 \rangle$ as a function of time and the results are shown in Fig. 27. For the potential of larger amplitude (curve 1 of Fig. 15), which fits the X-ray structure data with reasonable accuracy, they find $D = 1.9 \times 10^{-5}$ cm^2 sec^{-1}, while for the other potential (curve 2 of Fig. 15) they obtain the value 5.8×10^{-5} cm^2 sec^{-1}. These values are to be compared with the experimental result of 4.2×10^{-5} cm^2 sec^{-1} at 100°C.

The results of Fig. 27 show no sign of the extreme solid-like behaviour of $\langle r^2 \rangle$ reported by Randolph (1964) for $t > 10^{-13}$ sec. However, for $t < 10^{-13}$ sec, that is in the region of gas-like behaviour, there is fair agreement between these calculations and the experimental results of Randolph.

In spite of the absence of solid-like behaviour, there are suspicions of oscillations in $\langle r^2 \rangle$ in the curves shown in Fig. 27 and an interesting question for the future will be to ascertain whether, with oscillatory potentials extending well beyond the truncation point of 8 Å adopted by Paskin and Rahman, the behaviour found by Randolph could be reproduced. The indications are then, as referred to above, that the self-diffusion constant itself is going to afford a useful test of long-range potentials, as it appears quite sensitive to the detailed forms chosen. Further work in this area, both theoretical and experimental, is clearly called for.

8.8 Coherent scattering function

In spite of the fact that the calculation of the self part $G_s(\mathbf{r}, t)$, or $S_{\text{inc}}(\mathbf{K}, \omega)$, of the van Hove correlation function is still incomplete, it is certainly true that the coherent scattering function presents much greater difficulties. We shall therefore content ourselves with a brief discussion of an interesting physical approach due to Vineyard (1958), and then consider its connection with the small time expansion method which generalizes the approach of section 8.4.

8.8.1 VINEYARD'S CONVOLUTION APPROXIMATION

Writing the van Hove correlation function in the form

$$G(\mathbf{r}, t) = G_s(\mathbf{r}, t) + G_d(\mathbf{r}, t), \qquad (8.47)$$

we wish, if possible, to generate G_d from our knowledge of the self part, G_s.

Following Vineyard, we suppose that atom 1 is at the origin at time $t = 0$, and that simultaneously atom 2 is at $\mathbf{r}'$. Then there exists a certain probability, $H(\mathbf{r}, \mathbf{r}', t)$ say, that at a later time t atom 2 will have diffused into unit volume around $\mathbf{r}$, thereby suffering a displacement $\mathbf{r} - \mathbf{r}'$.

Then we can write for $G(\mathbf{r}, t)$ the equation

$$G(\mathbf{r}, t) = G_s(\mathbf{r}, t) + \varrho_0 \int g(\mathbf{r}') \, H(\mathbf{r}, \mathbf{r}', t) \, d\mathbf{r}'. \qquad (8.48)$$

The first term is, of course, simply the probability that the atom initially at the origin has migrated to $\mathbf{r}$ in the time t. In the second term, $g(\mathbf{r}')H(\mathbf{r}, \mathbf{r}', t) \, d\mathbf{r}'$ is the probability that any other atom is in the volume $d\mathbf{r}'$ at $\mathbf{r}'$ at time $t = 0$ and migrates to $\mathbf{r}$ in time t.

The essential approximation which is now made is to argue that, while $H(\mathbf{r}, \mathbf{r}', t)$ is actually dependent on the fact that we know an atom to be at the origin at time $t = 0$, it can be replaced to fair accuracy by the probability that an atom at $\mathbf{r}'$ will migrate to $\mathbf{r}$ in time t in the absence of knowledge about the positions of any other atoms. This latter probability is the self-correlation function G_s, and hence we replace (8.48) by

$$G(\mathbf{r}, t) \doteq G_s(\mathbf{r}, t) + \varrho_0 \int g(\mathbf{r}') \, G_s(\mathbf{r} - \mathbf{r}', t) \, d\mathbf{r}'. \qquad (8.49)$$

In terms of the intermediate scattering function $F(\mathbf{K}, t)$, we can write immediately from (8.49)

$$F(\mathbf{K}, t) \doteq F_s(\mathbf{K}, t) + \Gamma(\mathbf{K}) \, F_s(\mathbf{K}, t) \qquad (8.50)$$

where

$$\Gamma(\mathbf{K}) = \varrho_0 \int g(r) \, e^{i\mathbf{K} \cdot \mathbf{r}} \, d\mathbf{r}, \qquad (8.51)$$

which only differs trivially from the structure factor $S(\mathbf{K})$.

8.8.2 DELAYED CONVOLUTION APPROXIMATION FOR ARGON

From the molecular dynamics calculation, Rahman (1964) found that an approximation better than eqn. (8.50) is

$$F(\mathbf{K}, t) = F_s(\mathbf{K}, t) + \Gamma(\mathbf{K})F_s(\mathbf{K}, t'(t)) \qquad (8.52)$$

93

where $t'(t)$ is some delayed time which takes account of the correlation due to the presence of an atom at the origin initially. Using the spatial moments for liquid argon as calculated by Rahman from molecular dynamics and employing this delayed convolution approximation, Desai and Nelkin (1966) have calculated the energy distribution of scattered neutrons and $S(K, \omega)$ for liquid argon. They show that Rahman's computer experiments are in semiquantitative agreement with the inelastic neutron scattering experiments of Chen *et al.* (1965).

8.8.3 ANALYTIC STRUCTURE

It is of interest to note at this point that, as discussed in section 5, the structure factor $S(\mathbf{K})$, and hence $\Gamma(\mathbf{K})$, can have non-analytic behaviour at certain points in $\mathbf{K}$ space.

For example, for van der Waals forces, a K^3 term enters the small K expansion of $S(\mathbf{K})$, and hence from eqn. (8.50), will clearly be reflected in the pair part of $G(\mathbf{r}, t)$ according to the Vineyard approximation. It is an interesting problem as to whether the coefficient a_3 of eqn. (5) of Appendix 5 can be obtained as a function of time t (or ω).

More important for the present interests, the dielectric screening theory of a liquid metal suggests non-analyticity in the structure factor at $K = 2k_f$. It will be of interest to examine further the question of whether the Vineyard approximation will give the correct analytical structure of $S(\mathbf{K}, \omega)$.

8.8.4 SMALL TIME EXPANSION OF INTERMEDIATE SCATTERING FUNCTION $F(K, t)$

Because we can thereby make some contact between a presumably exact expansion and the physical approximations of Vineyard, we shall briefly summarize the results of the small time expansion of the intermediate scattering function, given for the self-correlation function in section 8.4. As Nijboer and Rahman

(1966) have shown, we can write with $\beta = (k_B T)^{-1}$,

$$F(\mathbf{K}, t) = e^{\frac{-K^2 t^2}{2\beta M}} \left\{ 1 + \Gamma(K) \left(1 + \frac{K^2 t^2}{2\beta M} + \frac{K^4 t^4}{8\beta^2 M^2} \right) \right.$$
$$\left. - \frac{K^2 t^4}{4!\,\beta M^2} \varrho_0 \int g(r) \left[\cos Kx - 1 \right] \frac{\partial^2 \phi}{\partial x^2} \, d\mathbf{r} \right\}. \quad (8.53)$$

This expansion can be compared with the result of the convolution approximation, which, from (8.50), (8.44) and (8.43), is

$$F_{\text{conv}}(\mathbf{K}, t) = e^{\frac{-K^2 t^2}{2\beta M}} [1 + \Gamma(K)] \left[1 + \frac{K^2 t^4 \varrho_0}{3.4!\,\beta M^2} \int g(r) \, \nabla^2 \phi(r) \, d\mathbf{r} + \dots \right].$$
$$(8.54)$$

8.9 Sum rules

In view of the difficulties we have seen exist in calculating precisely the van Hove correlation functions, certain exact sum rules have played a valuable role in the theory so far. These are collected together below, and the reader is referred to the paper by de Gennes (1959), for example, for a fuller discussion. It turns out that the convolution form (8.54) violates these sum rules and this, of course, urges caution in the use of such an approximation. The sum rules essentially give us the second and fourth moments of ω and are as follows:

$$\langle \omega_{\text{coh}}^2 \rangle = \frac{K^2 k_B T}{M S(K)} \quad (8.55)$$

$$\langle \omega_{\text{inc}}^2 \rangle = \frac{K^2 k_B T}{M} \quad (8.56)$$

$$\langle \omega_{\text{coh}}^4 \rangle = \frac{K^4 k_B T}{M^2 S(K)} \left[3 k_B T + \int d\mathbf{r} g(\mathbf{r}) \left\{ \frac{1 - \cos Kx}{K^2} \right\} \frac{\partial^2 \phi}{\partial x^2} \right]$$
$$(8.57)$$

and

$$\langle \omega_{\text{inc}}^4 \rangle = \frac{K^4 k_B T}{M^2} \left[3 k_B T + \int d\mathbf{r} g(\mathbf{r}) \frac{1}{K^2} \frac{\partial^2 \phi}{\partial x^2} \right]. \quad (8.58)$$

These sum rules are valid provided we can use velocity independent pair potentials. The fourth moments depend sensitively on the potential and therefore clearly need further study. As remarked above, it is not difficult now to show that the convolution approximation violates these sum rules, while the small time expansions developed earlier, of course, satisfy these requirements. Nevertheless, in giving us a picture of what is essentially going on, the convolution approximation remains appealing, in spite of its quantitative defects.

CHAPTER 9

Electron states

WE TURN finally to the aspect of the liquid metals problem on which least is known with certainty, namely the calculation of the electronic states in a liquid metal. Since this involves great theoretical complexity, it is natural to ask what experiment has to say, in setting up an appropriate theory. Unfortunately even the evidence afforded by experiment is far from unambiguous. Briefly relevant experimental data are available on at least six aspects of the problem:

  (i) D.C. transport properties
 (ii) A.C. conductivity
(iii) Pauli spin susceptibility
 (iv) Knight shift
  (v) Soft X-ray emission
 (vi) Positron annihilation.

The 'classical' solid state experiments, leading to information on the Fermi surface (de Haas–van Alphen effect, etc.) are, unfortunately, not applicable to liquids, as the electronic mean free path is too short.

As we discussed in Chapter 7, the d.c. transport properties, and in particular resistivity and thermopower, probably do not depend on the true density of states $n(E)$ in the liquid metal, but rather on the free electron density of states, $n_0(E)$. However, Mott (1966) has argued that if $n(E)$ gets very small, then it should enter the theory, and that this probably happens in liquid Hg. Furthermore, experimental evidence afforded by the Hall coefficient pointed to free-electron-like behaviour in the liquid. However, the free-electron value of the Hall coefficient R comes also from any theory in which the dispersion relation $E(k)$ and the relaxation time $\tau(k)$ are isotropic, which seems to be the case in the liquid state. Nevertheless, whereas $R = 1/ne$ for most liquid metals, a number of workers agree that for Pb, $R_{Pb} < 1/ne$. This may be due to a dependence on mean free path.

Turning to (ii) above, we saw in Chapter 7, §7.10 that while the Drude theory works well in representing the optical properties of most liquid metals, it does not hold for liquid Hg. This may well be closely linked with Mott's theory referred to above.

Regarding the Pauli susceptibility χ_p, band theory relates this directly to the density of states at the Fermi surface. χ_p, so far, has only been measured directly by electron spin resonance for liquid Li (Enderby, Titman and Wignall, 1964). It was found to increase by, at most, a few per cent through the melting point. Since the behaviour of lithium in the solid is not free-electron-like, the indications here are that the electron states are more closely related to those in the solid than to free electron theory. Connected with this, the change in Knight shift across the melting point has been found to be small (for simple metals, i.e. excluding Ga, Bi and Sb) except for Cd (Seymour and Styles, 1964). It is worth noting that Cd has also a large change in the Hall coefficient; in fact a change in sign (Enderby, 1963).

The indications from the available soft X-ray spectra, in particular for Al and Li, are that there are no very dramatic changes through the melting point. Especially for Li, this seems to confirm our conclusion above that the electron states in liquid Li are very like those in the solid state. Finally, positron annihilation results, which, taken at their face value, give us direct information on the momentum distribution of the conduction electrons (cf. Chapter 7), remain very difficult to interpret (see, however, Stewart *et al.*, 1963).

The experiments referred to above seem to the present writer to point towards the rather direct connection between the liquid and the solid. Nevertheless, while the Bloch wave functions are easy to classify, and also, in principle, easy to calculate, in the crystalline state, the nature of the eigenfunctions in a complex array of ions (one instantaneous ion configuration in the liquid, say) must be so complicated that it is not very useful to try to describe the wave functions in detail. However, it is known that there is often a strong tendency towards localization of wave functions in a disordered array (Gubanov, 1965; Makinson and Roberts, 1960).

We shall not go into any detail concerning the wave functions here, as either density matrix calculations or the Green function method seem much more appropriate. Edwards was the first person to draw attention to the power of the Green function approach and the remainder of this Chapter is based largely on his theory and its subsequent application. Unfortunately, for the reasons discussed above, we expect the following considerations to apply only when the electrons in the metallic crystalline state are not very different from free electrons; a severe limitation of course.

9.1 Green function calculation

Edwards (1962; see also Cusack (1963) for a review of this work) bases his arguments on partial summations of perturbation series, which are certainly far from rigorous. However, his theory has the considerable merit that progress can be made in calculating electronic energy states from a knowledge of the structure factor $S(K)$ and the single-centre scattering potential of a 'dressed' ion. Thus, within an admittedly very approximate framework, there is a hope of giving a theory of the electron states in terms of the same basic quantities that have occupied such a central role in our account of liquid metals in the earlier chapters.

As remarked above, we give up a direct attempt to study the individual electron wave functions, though we start out from the Schrödinger equation for an electron moving in the field of ions at positions $\mathbf{R}_i$, the ions being simulated by localized scattering potentials $V_i(\mathbf{r})$. Then we have

$$-\frac{\hbar^2}{2m}\nabla^2\Psi+\sum_i V_i(\mathbf{r}-\mathbf{R}_i)\,\Psi = -\frac{\hbar}{i}\frac{\partial\Psi}{\partial t}. \qquad (9.1)$$

We now effect a generalization of this equation to deal with the

99

Liquid Metals

Green functions G_+ and G_- which satisfy

$$\left[\frac{\hbar}{i}\frac{\partial}{\partial t}-\frac{\hbar^2}{2m}\nabla^2\pm i\varepsilon+\sum_i V_l(\mathbf{r}-\mathbf{R}_i)\right]G_{\pm}(\mathbf{r},\mathbf{r}';t,t')$$
$$=\delta(\mathbf{r}-\mathbf{r}')\,\delta(t-t'). \quad (9.2)$$

The small imaginary quantity $\pm i\varepsilon$ defines $G_{\pm}$ to be the solutions corresponding to incoming or to outgoing waves.

Then the important result we require is that the average density of states, $\langle n(E)\rangle$ say, when the ionic positions are distributed according to some prescribed information, is given in terms of G_+ and G_- by

$$\langle n(E)\rangle = \frac{\Omega}{(2\pi)^3}\int \varrho(E,\mathbf{k})\,d\mathbf{k}, \quad (9.3)$$

where $\varrho(E,\mathbf{k})$ is the Fourier transform of

$$\varrho(\mathbf{r}-\mathbf{r}',\,t-t') = \left(\frac{i}{2\pi}\right)[\langle G_+\rangle-\langle G_-\rangle]. \quad (9.4)$$

The averages of G_+ and G_- are simply functions of $\mathbf{r}-\mathbf{r}'$ since the liquid is spatially homogeneous.

The intermediate quantity $\varrho(E,\mathbf{k})$ connecting the density of states with the Green functions is the probability of finding an electron with energy E and momentum $\mathbf{k}$.

9.2 Perturbation theory

Following Edwards, we now find it convenient to define a 'perturbation' $u(\mathbf{r})$ as the potential energy in (9.2) minus its mean value, that is

$$u(\mathbf{r}) = \sum_i V_l(\mathbf{r}-\mathbf{R}_i)-\sum_i \langle V_l(\mathbf{r}-\mathbf{R}_i)\rangle. \quad (9.5)$$

By shifting the origin of energy appropriately, we can expand G in terms of u. There is then no difficulty in averaging the resulting series, which can be written formally as

$$G_+ = G_0-G_0\langle u\rangle\, G_0+G_0\langle uG_0u\rangle\, G_0-\ldots \quad (9.6)$$

where G_0 satisfies the free-particle equation

$$\left[\frac{\hbar}{i}\frac{\partial}{\partial t}-\frac{\hbar^2}{2m}\nabla^2+i\varepsilon\right]G_0 = \delta(\mathbf{r}-\mathbf{r}')\,\delta(t-t'). \quad (9.7)$$

100

The solution of this, in terms of E and $\mathbf{k}$ may be written down immediately as

$$\left[E - \left(\frac{\hbar^2}{2m}\right) k^2 + i\varepsilon \right] G_0 = 1. \tag{9.8}$$

Now we have chosen the average value of u by (9.5) to be zero and so the second term on the right-hand-side of (9.6) vanishes. Edwards further argues that all subsequent odd powers of u are of the same order in the density as the even terms which precede them, whereas, of course, they are one order higher in the potential $u(\mathbf{r})$. Thus, these terms are neglected and only the even terms in u are considered. The achievement of Edwards is then to show that, under a rather wide class of conditions, the resulting series can be summed.

9.3 Partial summation of Green function series

The term in u^2 in (9.6) means explicitly

$$\int G_0(\mathbf{r}, \mathbf{r}'') \langle u(\mathbf{r}'') G_0(\mathbf{r}'', \mathbf{r}''') u(\mathbf{r}''') \rangle G_0(\mathbf{r}''', \mathbf{r}') \, d\mathbf{r}'' \, d\mathbf{r}'''$$

$$\tag{9.9}$$

and hence, from (9.5), involves the average

$$\sum_{i,j} \langle V_l(\mathbf{r}_1 - \mathbf{R}_i) \, V_l(\mathbf{r}_2 - \mathbf{R}_j)$$

$$= (2\pi)^{-6} \int \tilde{V}(\mathbf{k}) \, \tilde{V}(\mathbf{j}) \, \langle \exp(i\mathbf{k} \cdot \mathbf{R}_i + i\mathbf{j} \cdot \mathbf{R}_j) \rangle \tag{9.10}$$

$$\times \exp(-i\mathbf{k} \cdot \mathbf{r}_1 - i\mathbf{j} \cdot \mathbf{r}_2) \, d\mathbf{r}_1 \, d\mathbf{r}_2$$

where

$$\tilde{V}(\mathbf{k}) = \int V_l(\mathbf{r}) \, e^{i\mathbf{k} \cdot \mathbf{r}} \, d\mathbf{r}. \tag{9.11}$$

Performing the averaging and using the homogeneity of the fluid, we can write

$$\left\langle \sum_{i,j} \exp i(\mathbf{k} \cdot \mathbf{R}_i + \mathbf{j} \cdot \mathbf{R}_j) \right\rangle$$

$$= (2\pi)^3 \, \Omega^{-1} \sum_{i,j} \langle \exp i\mathbf{k} \cdot (\mathbf{R}_i - \mathbf{R}_j) \rangle \, \delta(\mathbf{k} + \mathbf{j})$$

$$= (2\pi)^3 \, \Omega^{-1} \delta(\mathbf{k} + \mathbf{j}) \, S(\mathbf{k}), \tag{9.12}$$

where we have simply used the definition of the structure factor $S(\mathbf{k})$. Thus, combining (9.10) and (9.12) we find

$$\left\langle \sum_{i,j} V_i(\mathbf{r}_1 - \mathbf{R}_i) \, V_i(\mathbf{r}_2 - \mathbf{R}_j) \right\rangle$$
$$= (2\pi)^{-3} \Omega^{-1} \int |\tilde{V}(\mathbf{k})|^2 \, S(\mathbf{k}) \exp\{i\mathbf{k} \cdot (\mathbf{r}_1 - \mathbf{r}_2)\} \, d\mathbf{k}. \quad (9.13)$$

If we now return to (9.9) and take its Fourier transform, we find the result

$$G_0(\mathbf{k}) \, \Sigma(\mathbf{k}) \, G_0(\mathbf{k})$$

where

$$\Sigma(\mathbf{k}) = \Omega^{-1} \int d\mathbf{j} S(\mathbf{k} - \mathbf{j}) \, G_0(\mathbf{j}) \, |\tilde{V}(\mathbf{k} - \mathbf{j})|^2. \quad (9.14)$$

This is, in fact, the basic result, as we shall see below. For if we now consider the term of order u^4 in (9.6) in an exactly similar way, then it can be written in the form

$$\Sigma(\mathbf{k}) \, G_0(\mathbf{k}) \, \Sigma(\mathbf{k}) + \Sigma_4(\mathbf{k}), \quad (9.15)$$

where, as Edwards argues, $\Sigma_4(\mathbf{k})$ is small compared with the first term in (9.15) in a wide variety of situations of physical interest in liquid metals. We shall not go into details here, but simply note that, in this approximation, we may write (9.6) in the form

$$\langle G \rangle = G_0 + G_0 \Sigma G_0 + G_0 \Sigma G_0 \Sigma G_0 + \ldots \quad (9.16)$$

This is a geometric series and we can write the solution in the form

$$\left[E - \frac{\hbar^2 k^2}{2m} + i\varepsilon - \Sigma(\mathbf{k}) \right] \langle G \rangle = 1. \quad (9.17)$$

This is the central result, and Σ can be calculated from (9.14) using the measured structure factor $S(\mathbf{k})$, once a choice of scattering potential has been made. We turn immediately to discuss the results which have been obtained using this method.

9.4 Results

Several workers have by now applied the Edwards theory to specific liquid metals. Thus, Watabe and Tanaka (1964) have obtained results for both monovalent and polyvalent metals as considered below. In some cases, however, very crude representations of $\tilde{V}(\mathbf{k})$ were employed. A calculation has also been carried out in liquid lithium (Jones and Stott; private communication) and while a pseudopotential method has been employed, it seems from comparison with the solid (Stott and March, 1966) that the nearly free electron approximation is not good and the assumptions of the Edwards treatment may not be valid. However, these workers all agree that the corrections to the free electron results according to the Edwards theory, are not major for monovalent metals.

Watabe and Tanaka calculate the dispersion relation $E(\mathbf{k})$ (assuming implicitly that it is not blurred out markedly by the disorder scattering: cf. remarks below) by noting from (9.17) that

$$E(\mathbf{k}) = \frac{\hbar^2 k^2}{2m} + Re\,\Sigma\,[\mathbf{k}, E(\mathbf{k})]. \tag{9.18}$$

In fact, the imaginary part of Σ gives the level blurring, which, while related to the electronic mean free paths in liquid metals, we shall choose to neglect as a first approximation.

Writing (9.18) in units of the Fermi energy $\dfrac{\hbar^2 k_f^2}{2m}$ and the Fermi momentum k_f, we obtain from (9.14) and (9.8),

$$Re\,\frac{\Sigma(k, \zeta)}{(\hbar^2 k_f^2/2m)} = \frac{3}{2k}\int_0^{\infty} K\left|\frac{U(K)}{(\hbar^2 k_f^2/2m)}\right|^2 ZS(\mathbf{K})\ln\left|\frac{\zeta-(2K-k)^2}{\zeta-(2K+k)^2}\right| dK \tag{9.19}$$

where $K = \dfrac{|\mathbf{K}|}{2k_f}$, Z as usual is the valency and $U(K) = N\Omega^{-1}\tilde{V}(K)$.

By substituting in measured values for $S(K)$ with a chosen potential $U(K)$, we can proceed to solve (9.18) numerically, to find $E(\mathbf{k})$.

The density of states can be found from (9.3) and if again we neglect the imaginary part of Σ then we obtain

$$\langle n(E) \rangle = 2\sum_{\mathbf{k}} \delta[E - E(\mathbf{k})]\left\{\left[1 - \frac{\partial Re\,\Sigma(k, \zeta)}{\partial \zeta}\right]_{\zeta = E(\mathbf{k})}\right\}^{-1}. \tag{9.20}$$

Liquid Metals

9.4.1 ALKALI METALS

For the alkali metals, Watabe and Tanaka have carried out calculations with a crude form of $U(K)$ given by

$$U_{\text{bare}}(K) = \frac{-N}{\Omega} \frac{4\pi Z e^2}{K^2} \cos aK. \tag{9.21}$$

The factor $\cos aK$ comes from the assumption that the pseudopotential may be taken to be zero in the core region, of radius a, and to be purely Coulombic outside. Choosing a so that the band gap in the solid is correctly given, and screening $U_{\text{bare}}(K)$ in the manner discussed in Chapter 7 to obtain $U(K)$, we can proceed to calculate $E(\mathbf{k})$ and $\langle n(E) \rangle$.

For liquid Na, the factor $\cos aK$ makes $U(K)$ negligible near $K = 2k_f$ and $|U(K)|^2 S(K)$ becomes very small in the whole range

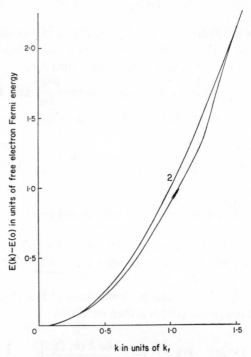

FIG. 28. $E(\mathbf{k})$ for liquid Rb (1) Liquid (2) Free electron result

of integration. Therefore $E(k)$ is practically unchanged from the free electron spectrum.

For liquid Rb, calculations were also carried out, with a pseudo-potential $U(K)$ given essentially by the rough approximation of eqn. (7.41) which we write in the dimensionless form

$$\left| \frac{U(K)}{\hbar^2 k_f^2/2m} \right| = \frac{2}{3}\left(\frac{\alpha r_s}{\pi a_0}\right)\frac{1}{K^2+q_s^2} \qquad (9.22)$$

where $\alpha = \left(\frac{4}{9\pi}\right)^{\frac{1}{3}}$, and $q_s^2 = \frac{\alpha r_s}{\pi}$. r_s is the mean interelectronic spacing defined by $n_0 = 3/4\pi r_s^3$ and taking $r_s = 5.2$ atomic units the $E(k)$ relation is shown in Fig. 28.

In this case the pseudopotential around $2k_f$, as given by equation (9.22), is not small. This leads to a small bump in $E(k)$ near $2k_f$, that is in the region of the peak of the structure factor. However, the correction to the density of states is still only a few per cent at the Fermi surface.

9.4.2 POLYVALENT METALS

We shall conclude with a brief summary of the results of Watabe and Tanaka (1964) for polyvalent metals. Using the same crude form (9.22), with $r_s = 2.30$ atomic units and the data of Gamertsfelder (1941) for $S(K)$, the dispersion relation $E(k)$ has been calculated for liquid zinc. Because of the structure in $S(K)$, there are two anomalous regions in $E(k)$, as shown in Fig. 29. Some doubts must exist as to the accuracy of this rather old X-ray data for $S(K)$, but, nevertheless, employing this $E(k)$ relation to calculate the density of states, a very different result emerges at the Fermi energy, as shown in Fig. 30.

Watabe and Tanaka interpret in a similar way the soft X-ray L_{23} emission spectrum from liquid Al, measured by Catterall and Trotter (1963). We feel that, while their interpretation of this is plausible, there are so many approximations in the theory to date that it is perhaps premature to conclude that we have any reliable picture of the density of states in a polyvalent metal. Clearly, however, this is a field of great interest and one where the essential foundations have been laid for what should be substantial progress over the next decade.

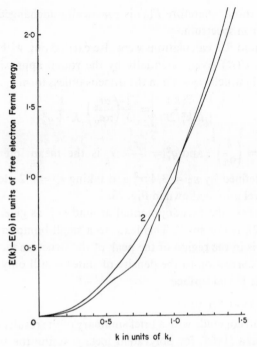

FIG. 29. $E(k)$ for liquid Zn. (1) Liquid (2) Free electron result

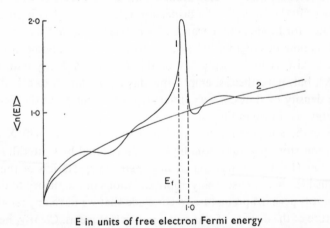

FIG. 30. Density of states $\langle n(E) \rangle$ for liquid Zn in units of free electron density of states at Fermi energy $(\hbar^2 k_f^2 / 2m)$
(1) Liquid (2) Free electron result

APPENDIX 1

Long-wavelength limit of structure factor

DETAILED discussion of the density fluctuations argument leading to the relation between the long wavelength limit of the structure factor and the isothermal compressibility are available elsewhere (cf. Landau and Lifshitz, 1958). We shall consider here briefly an explicit model, introduced by Feynman and Cohen (1956) which makes the result plausible.

To understand the behaviour of $S(k)$ for small k, we first anticipate that when we are concerned solely with long wavelength disturbances, the liquid can be treated as a continuous compressible medium. If $\varrho(\mathbf{r}, t)$ is the number of atoms per unit volume in such a medium, we define the normal coordinates ϱ_k by

$$\varrho_k = \int \varrho(\mathbf{r}, t)\, e^{i\mathbf{k}\cdot\mathbf{r}}\, d\mathbf{r} \qquad (A1.1)$$

and then the energy is, assuming the ϱ_k vary harmonically,

$$E = \frac{1}{2} \sum_k \left(\frac{m}{Nk^2}\right) [\dot\varrho_k \dot\varrho_k^* + \omega_k^2 \varrho_k \varrho_k^*] \qquad (A1.2)$$

where $\omega_k = ck$ and c is the velocity of sound.

Passing to quantum mechanics, $\varrho(\mathbf{r})$ is replaced by the operator $\sum_i \delta(\mathbf{r} - \mathbf{r}_i)$ and ϱ_k then goes over into the quantum mechanical normal coordinate

$$\varrho_k = \sum_i \exp(i\mathbf{k}\cdot\mathbf{r}_i). \qquad (A1.3)$$

Appendix 1

The structure factor $S(k)$ is, from (7.39), the expectation value of $N^{-1} |\varrho_k|^2$. Since the average values of the kinetic and potential energies are equal, from the virial theorem, for a harmonic oscillator, it follows from (A1.2) that

$$S(k) = \frac{\langle E_k \rangle}{mc^2}, \tag{A1.4}$$

where $\langle E_k \rangle$ is the average energy of the oscillator representing sound of wave number $\mathbf{k}$. When $T = 0°K$, all the oscillators are in their ground states, and hence

$$\langle E_k \rangle = \tfrac{1}{2} \hbar \omega_k = \tfrac{1}{2} \hbar c k. \tag{A1.5}$$

Thus, it follows from equations (A1.4) and (A1.5) that

$$S(k) = \frac{\hbar k}{2mc}, \tag{A1.6}$$

which is true for the quantal fluid He^4 at $T = 0°K$ and small k.

Our prime interest is in the elevated temperature result, and since, in this case, the probability of finding the oscillator representing phonons of wave number $\mathbf{k}$ in its nth excited state is proportional to $\exp(-E_n/k_B T)$, it follows that

$$S(k) = \left(\frac{\hbar k}{2mc} \right) \coth \frac{1}{2} \beta \hbar c k$$

$$= (\beta mc^2)^{-1} + \frac{\beta \hbar^2}{12m} k^2 + \dots . \tag{A1.7}$$

Here $\beta = (k_B T)^{-1}$ and as $k \to 0$ we find

$$S(0) = \varrho_0 k_B T K_T \tag{A1.8}$$

the desired result.

Dielectric function of high density Fermi gas

WE SHALL briefly summarize here the method of solution of the basic equation (3.25) describing the shielding of a perturbation according to the wave theory. Using atomic units we have

$$\nabla^2 V = \frac{2k_f^2}{\pi^2} \int d\mathbf{r}' \, V(\mathbf{r}') \frac{j_1(2k_f |\mathbf{r}-\mathbf{r}'|)}{|\mathbf{r}-\mathbf{r}'|^2}. \qquad (A2.1)$$

Introducing the Fourier transform $\tilde{V}(\mathbf{k})$ of the potential through eqn. (3.28) and taking the charge being shielded as unity for convenience, the Fourier transforms of the pure and screened Coulomb potentials are simply $-4\pi/k^2$ and $-4\pi/(k^2+q^2)$, where, from (3.12), $q^2 = 4k_f/\pi$. In the wave theory, q^2 is replaced by a quantity intimately related to the Fourier transform of the wave factor $j_1(2k_f r)/r^2$ appearing on the right-hand side of (A2.1), as we now show.

Substituting eqn. (3.28) into (A.2.1) we find immediately

$$\int (-k^2) \, \tilde{V}(\mathbf{k}) \, e^{i\mathbf{k}\cdot\mathbf{r}} \, d\mathbf{k} = \frac{2k_f^2}{\pi^2} \int \tilde{V}(\mathbf{k}) \, e^{i\mathbf{k}\cdot\mathbf{r}'} \frac{j_1(2k_f |\mathbf{r}-\mathbf{r}'|}{|\mathbf{r}-\mathbf{r}'|^2} \, d\mathbf{r}' \, d\mathbf{k}$$

$$+ 4\pi \int e^{i\mathbf{k}\cdot\mathbf{r}} \, d\mathbf{k} \qquad (A2.2)$$

where the second term on the right-hand side of (A2.2) accounts explicitly for the presence of unit point charge at the origin (charge density equal to a δ function, with constant Fourier components therefore). But the integral over $\mathbf{r}'$ required in the second

Appendix 2

term can now be performed by writing $\mathbf{R} = \mathbf{r}' - \mathbf{r}$, and then we have

$$\int (-k^2)\, \tilde{V}(\mathbf{k})\, e^{i\mathbf{k}\cdot\mathbf{r}}\, d\mathbf{k} = \int \tilde{V}(\mathbf{k})\, e^{i\mathbf{k}\cdot\mathbf{r}}\, J(k_f, k)\, d\mathbf{k} + 4\pi \int e^{i\mathbf{k}\cdot\mathbf{r}}\, d\mathbf{k},$$

(A2.3)

where $J(k_f, k)$ is evidently given by

$$J(k_f, k) = \frac{2k_f^2}{\pi^2} \int e^{i\mathbf{k}\cdot\mathbf{R}}\, \frac{j_1(2k_f R)}{R^2}\, d\mathbf{R}.$$

(A2.4)

This definite integral can be evaluated explicitly and after some calculation we find

$$J(k_f, k) = \left[\frac{2k_f}{\pi} + \left(\frac{2k_f^2}{\pi k} - \frac{k}{2\pi} \right) \ln \left| \frac{k + 2k_f}{k - 2k_f} \right| \right].$$

(A2.5)

From (A2.3) we have then the desired solution

$$\tilde{V}(\mathbf{k}) = \frac{-4\pi}{k^2 + J(k_f, k)}.$$

(A2.6)

Obviously, the role of q^2 in the screened Coulomb Fourier components is taken over by J, with its characteristic singularity at $k = 2k_f$ from (A2.5). The connection between the two treatments is readily seen by taking the long wavelength limit $(k \to 0)$ in (A2.5), when we find

$$J(k_f, 0) = \frac{4k_f}{\pi} = q^2.$$

(A2.7)

Using the definition of the dielectric function in eqn. (3.32), the result (3.34) follows immediately from (A2.4) and (A2.6).

The argument presented above, as the work of Langer and Vosko (1959) shows, is only valid rigorously in a very high density Fermi gas, where the mean interelectronic spacing is very much less than an atomic unit (Bohr radius $\hbar^2/me^2$). In liquid metals, the mean spacing is between two and six atomic units and the theory is being stretched outside its proper range of validity. We must expect that exchange and correlation corrections will have some significance, and these have been considered by various workers. We shall not go into further detail here.

Electrostatic model of ion–ion interaction in Fermi gas

WE WILL discuss here the so-called electrostatic model for the interaction energy between charged centres in the Fermi gas. We shall use a semi-classical argument here (Alfred and March, 1957) for simplicity. The essential result is valid in a full wave-mechanical treatment (Corless and March, 1961).

Let us suppose that ions 1 and 2, embedded in the bath of conduction electrons, are screened such that the potential due to ion 1 alone is V_1 and that due to 2 is V_2. Then in the linear (Born) approximation of Chapter 3, it is clear that the total potential V is just the superposition potential, given by

$$V = V_1 + V_2. \tag{A3.1}$$

Hence each ion is surrounded by its own displaced charge, and this is not affected by bringing up further ions. This is the reason why the total potential energy of the ions can be written as a sum of pair potentials. It is a central assumption in this book, though it is of course recognized to involve approximations.

The interaction energy between the ions, separated by a distance r say, may now be obtained directly by calculating the difference between the total energy of the metal when the ions are separated by an infinite distance (that is neglecting surface effects by taking the limit as the volume tends to infinity) and when they are brought up to the mutual separation r. Clearly this is all to be done in the Fermi sea of constant density, and the result for the

Appendix 3

interaction energy will depend on the density, or equivalently, from eqn. (3.3), on the Fermi energy.

Let us consider the changes in the kinetic and potential energy separately. The act of introducing an ion carrying a charge Ze into the Fermi gas changes the kinetic energy in the following way. The mean energy of an electron in the unperturbed Fermi gas, T_0/N say, is entirely kinetic when we neglect exchange and correlation interactions and is given by

$$\frac{T_0}{N} = \int_0^{p_f} \frac{p^2}{2m} \left\{ \frac{4\pi p^2 \, dp}{(4/3)\pi p_f^3} \right\} \tag{A3.2}$$

since the probability of finding an electron with momentum between p and $p+dp$ is simply the factor in brackets in eqn. (A3.2). Thus

$$\frac{T_0}{N} = \frac{3}{10m} p_f^2 \tag{A3.3}$$

or $3/5$ of the Fermi energy E_f. The kinetic energy per unit volume is therefore $(3/5)E_f n_0$, and using eqn. (3.3) we find for the total kinetic energy

$$T_0 = \frac{3h^2}{10m} \left(\frac{3}{8\pi} \right)^{\frac{2}{3}} \int d\mathbf{r} \, n_0^{\frac{5}{3}}. \tag{A3.4}$$

We have written the equation in this form since it is now clear that, if we again follow our earlier procedure and apply free electron relations locally, we find for the kinetic energy of the Fermi gas, when perturbed by an ion

$$T = \frac{3h^2}{10m} \left(\frac{3}{8\pi} \right)^{\frac{2}{3}} \int d\mathbf{r} \, \{n(r)\}^{\frac{5}{3}}. \tag{A3.5}$$

Obviously if we are dealing with a very large volume Ω, then T becomes large with Ω, and therefore it is convenient to measure kinetic energy changes from the unperturbed Fermi gas state. Writing $(n-n_0) = \Delta n$, and assuming consistently with our first-order theory that Δn is small, we find

$$T-T_0 = E_f \int \Delta n \, d\mathbf{r} + \frac{E_f}{3n_0} \int d\mathbf{r} (\Delta n)^2 + 0(\Delta n^3) \tag{A3.6}$$

where we have used eqn. (3.3) to eliminate n_0 in favour of E_f in the first term. We now observe that the first term involves simply the normalization condition for the displaced charge, or in other words, the condition that the ionic charge Ze is screened completely by the electron distribution. Clearly, this term will make no contribution to the energy difference between infinitely separated ions and the ion pair at separation r.

We are now in a position to compute the changes in both kinetic and potential energies when we bring the ions together. We may write down the following contributions:

(i) The interaction energy between the charge Ze of one impurity and the perturbing potential, say V_2, due to the other.

(ii) The interaction energy between the displaced charge $\dfrac{q^2}{4\pi e^2} V_1$, round the first ion and the potential V_2 due to the other.

(iii) The change in kinetic energy.

These three terms are evidently given by:

(i) $\quad Z^2 e^2 \exp(-qr)/r \qquad\qquad\qquad\qquad\qquad$ (A3.7)

(ii) $\quad -\dfrac{q^2}{4\pi e^2} \displaystyle\int d\mathbf{r}\, V_1 V_2 \qquad\qquad\qquad\qquad$ (A3.8)

(iii) $\quad \left(-\dfrac{q^2}{4\pi e^2}\right)^2 \dfrac{E_f}{3n_0}\left[\displaystyle\int d\mathbf{r}\,\{(V_1+V_2)^2 - V_1^2 - V_2^2\}\right]$

$\qquad = \dfrac{q^2}{4\pi e^2} \displaystyle\int d\mathbf{r}\, V_1 V_2, \qquad\qquad\qquad\qquad$ (A3.9)

where (iii) follows from eqns. (3.10), (A3.1) and (A3.6). Thus, the contribution (ii) is just cancelled by the change in kinetic energy, and we are left with the final result

$$\Delta E = Z^2 e^2 \exp(-qr)/r. \qquad\qquad (A3.10)$$

But this is simply the electrostatic energy of an ion of charge Ze, sitting in the electrostatic potential $(Ze/r)\exp(-qr)$ of the second ion. Before the argument presented above was given by Alfred and March (1957), Lazarus (1954) had anticipated that such a result should hold on physical grounds, in the context of a discussion on

Appendix 3

impurity diffusion in metals. This electrostatic model, we stress, is not precise unless we restrict ourselves to the linear theory given above.

Equation (A3.10) is the basic result for the pair potential $\phi(r)$ between ions, in our model, and its form is shown in Fig. 8, curve b. While the region inside the ion core is clearly different (much more strongly repulsive) for finite ions, we are forced to reach the dismal conclusion that, with no minimum in $\phi(r)$, the liquid metal would be unstable. Even if we introduce core electrons, and argue that their dynamic motions will polarize the electrons in a second ion, such van der Waals interactions must again take place through the polarizable Fermi gas, and hence will be screened out in a distance of order q^{-1}. This statement is a little oversimplified. Strictly we should work with the dielectric function $\varepsilon(k)$ of section 3.4, generalized to include frequency dependence, i.e. $\varepsilon(k, \omega)$.

APPENDIX 4

Structure factor and direct correlation function for hard spheres

WE SHALL briefly summarize here the present state of our knowledge of the structure factor $S(K)$ and the Fourier transform $\tilde{f}(K)$ of the Ornstein–Zernike direct correlation function for a classical hard-sphere fluid.

The only exact results as yet available are for the leading terms in the virial expansion. In **r** space, the calculations were carried out by Nijboer and van Hove (1952) and the corresponding results in **K** space have recently been obtained by Ashcroft and March (1967).

The other point worthy of note is that for this system the Percus–Yevick equation (4.11) has an exact solution, derived independently by Wertheim (1963) and Thiele (1963). In accordance with expectation from the intimate relation between $f(r)$ and the pair potential $\phi(r)$ from the Percus–Yevick equation, the direct correlation function becomes zero outside the hard-sphere diameter. Study of the virial expansion suggests that $f(r)$ is a polynomial of order r^3 inside the hard sphere, and this can again be verified by substitution in the exact equation.

If we take R as the hard-sphere diameter, and introduce dimensionless variables

$$\eta = \tfrac{1}{6}\pi R^3 \varrho_0 \tag{A4.1}$$

and

$$x = \frac{r}{R} \qquad (A4.2)$$

then the direct correlation function $f(x)$ coincides with the form suggested by the behaviour of the virial coefficients, namely

$$f(x) = \alpha + \beta x + \gamma x^3; \quad x < 1 \qquad (A4.3)$$
$$= 0 \qquad\qquad\quad x > 1.$$

Here α, β and γ are functions of the packing density η, which from equation (A4.1) is seen to be the fraction of total fluid volume occupied by hard spheres. The defining equations for α, β and γ are in fact

$$\left.\begin{array}{l} (1-\eta)^4 \alpha = -(1+2\eta)^2 \\ (1-\eta)^4 \beta = 6\eta(1+2\eta)^2 \\ (1-\eta)^4 \gamma = -\tfrac{1}{2}\eta(1+2\eta)^2 \end{array}\right\}. \qquad (A4.4)$$

If we use equation (A4.3) in the density fluctuation formula (A1.8), then we readily find the equation of state

$$\frac{p}{\varrho_0 k_B T} = (1-\eta)^{-3}[1+\eta+\eta^2]. \qquad (A4.5)$$

As we remarked above, the only exact results with which to compare the Percus–Yevick result (A4.3) come from the virial expansion. A selection of the values for $S(K)$ obtained by Ashcroft and March is shown in Fig. A.1 for $\varrho_0 R^3 = 0.4$. The upper curve shows the results obtained from equation (A4.3), while the exact virial expansion leads to curve (a) below. Expanding the Percus–Yevick solution in powers of ϱ_0 yields, for the density used in Fig. A.1, a result graphically indistinguishable from the exact virial form. The results obtained from the virial expansion of the Born–Green equation (4.2) and the hyperchain equation (4.10) are also shown in curves (b) and (c) respectively. The Percus–Yevick method is evidently the most satisfactory of the three approximate theories for hard spheres at this density, though all theories agree at lower densities.

We can also calculate $\tilde{f}(K)/\tilde{f}(0)$ from the virial expansion, and for a density appropriate to the measurements of Henshaw (1957; see also Enderby and March, 1965) on liquid A, the results are

116

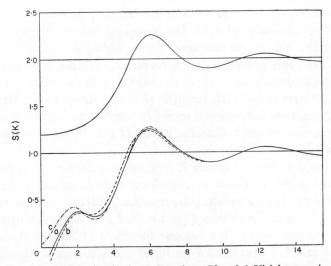

FIG. A.1. $S(K)$ for hard spheres. Density $\varrho_0 R^3 = 0.4$. Virial expansion
(a) Exact (b) Born–Green (c) Hyperchain
Upper curve. Exact Percus–Yevick result

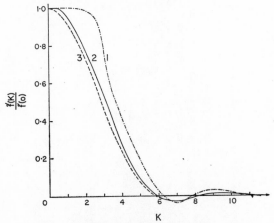

FIG. A.2. $\dfrac{\tilde{f}(K)}{\tilde{f}(0)}$ for hard spheres and for argon

(1) Experimental results
(2) Percus–Yevick theory (without density expansion)
(3) Exact (virial expansion) results

shown in Fig. A.2, together with the Percus–Yevick result (A4.3) and the experimental data. The deviations between theory and experiment are to be accounted for by adding a suitable tail to the hard core potential, and, for small K, Ashcroft and March conclude that the data are not inconsistent with the magnitude of the K^3 term in $S(K)$ (cf. equation (A5.6) of Appendix 5). Much more accurate experimental results for small angle scattering will be required before a quantitative test of that theory is possible however.

In eqn. (7.40) of Chapter 7, Ashcroft and Lekner (1966) have used the Percus–Yevick solution for $S(K)$ in calculations of the resistivity of liquid metals. This procedure yields quite good results because the first peak in $S(K)$ is described reasonably well by the hard sphere model. The peculiar behaviour of $S(K)$ at $2k_f$ and the form of $S(K)$ at small K for liquid metals cannot be correctly accounted for by such a model, however. As we said, this is not serious for resistivity calculations.

Asymptotic relation between total and direct correlation functions for van der Waals fluids

---

A5.1 Diagrammatic methods

We have stressed that available evidence points strongly to the asymptotic relation

$$f(r) \sim \frac{-\phi(r)}{kT}, \qquad (A5.1)$$

provided $h^2(r) \ll \left| \dfrac{\phi(r)}{kT} \right|$. This should be true away from the critical point. For van der Waals forces we have

$$\phi(r) \sim -Ar^{-6} \qquad (A5.2)$$

and hence

$$f(r) \sim \frac{A}{kT} r^{-6}. \qquad (A5.3)$$

As Enderby, Gaskell and March (1965) have shown, we can make use of the result (see, for example, Lighthill, 1958) that

$$\int_0^\infty F(x) \sin xr \, dx \sim \frac{F(0)}{r} - \frac{F''(0)}{r^3} + \frac{F^{IV}(0)}{r^5} - \cdots \qquad (A5.4)$$

To apply this result to the calculation of $f(r)$, we see from eqn. (2.22) that we must choose $F(K) \equiv K \dfrac{S-1}{S}$, and also we must

Appendix 5

have $F''(0) = 0$, $F^{IV}(0) \neq 0$. This immediately implies an expansion of $S(K)$ for small K of the form

$$S(K) = S(0) + a_2 K^2 + a_3 K^3 + \ldots \qquad (A5.5)$$

where a_3 is given by

$$a_3 = \pi^2 \varrho_0 \{S(0)\}^2 A/12kT. \qquad (A5.6)$$

Thus, $S(K)$ is not analytic at the origin, due to the van der Waals interaction. It follows after some calculation that

$$h(r) \sim \frac{1}{kT} \{S(0)\}^2 A r^{-6}$$

$$\sim \{S(0)\}^2 f(r). \qquad (A5.7)$$

Since $\{S(0)\}^2$ is usually small, $h(r) \ll f(r)$ for large r according to this argument.

A5.2 Born–Green theory

Unfortunately, as referred to in the text, the Born–Green theory does not lead to the asymptotic relation (A5.1) and we regard that as a defect of this approximation. The argument based on small K expansions of the form of (A5.5) has recently been extended by Gaskell (1965) to yield the relation in this approximation which replaces (A5.1) for van der Waals interactions. We shall summarize his argument below.

Returning to the Born–Green equation (4.2), we define the Fourier transform $\tilde{E}(K)$ of $E(r)$ through

$$E(r) = \int_r^\infty dt\, g(t)\, \frac{\phi'(t)}{kT} = \frac{1}{8\pi^3 \varrho_0} \int d\mathbf{K}\, \tilde{E}(\mathbf{K})\, e^{i\mathbf{K} \cdot \mathbf{r}}. \qquad (A5.8)$$

Asymptotically we must have

$$E(r) \sim -\frac{\phi(r)}{kT} \sim \frac{A r^{-6}}{kT} \qquad (A5.9)$$

and we can immediately see from the results (A5.6) and (A5.7) that the coefficient of K^3 in the small K expansion of $\tilde{E}(K)$ is $\pi^2 \varrho_0 A/12kT$. Hence we may write, in complete analogy with (A5.5),

$$\tilde{E}(K) = \tilde{E}(0) + b_2 K^2 + b_3 K^3 + \ldots \qquad (A5.10)$$

120

$\tilde{E}(0)$ may now be obtained from (A5.8) and it is readily shown that it has the form

$$\tilde{E}(0) = 4\pi\varrho_0 \left[\frac{r^3}{3} \int_r^\infty dt \, g(t) \, \frac{\phi'(t)}{kT}\right]_0^\infty + \frac{4\pi\varrho_0}{3} \int_0^\infty dr r^3 g(r) \frac{\phi'(r)}{kT}.$$

(A5.11)

Assuming $\phi(r)$ is such that the first term vanishes, and using the virial theorem (4.13) for the pressure p on the fluid, we find

$$\tilde{E}(0) = \frac{2}{\varrho_0 kT} (\varrho_0 kT - p).$$

(A5.12)

We wish now to obtain a new expression for a_3 in (A5.5), to replace the result (A5.6) of the diagrammatic methods. We can invert the above argument and note that $h(r)$ has an asymptotic expansion in inverse powers of r, the leading term being $(12a_3/\pi^2\varrho_0) \, r^{-6}$. For large r, we can therefore write the Born–Green equation in the form

$$\left(\frac{12a_3}{\pi^2\varrho_0}\right) r^{-6} - \frac{Ar^{-6}}{kT} = \frac{1}{8\pi^3\varrho_0} \int d\mathbf{K} \, e^{i\mathbf{K}\cdot\mathbf{r}} \{S(\mathbf{K}) - 1\} \tilde{E}(\mathbf{K}).$$

(A5.13)

The leading terms on the right-hand side of this equation vary also as r^{-6} and are determined by the coefficient of K^3 in the expansion of $\{S(\mathbf{K}) - 1\} \tilde{E}(\mathbf{K})$. Hence using (A5.10), (A5.12) and (A5.13) we find

$$a_3 = \frac{\pi^2\varrho_0^2 S(0) \, A}{12(2p - \varrho_0 kT)}$$

(A5.14)

which is obviously different from (A5.6). It follows that asymptotically

$$f(r) \sim \frac{-\phi(r)}{kT} \frac{\varrho_0 kT}{S(0)[2p - \varrho_0 kT]}.$$

(A5.15)

Putting in typical values for argon, the factor multiplying $\dfrac{-\phi(r)}{kT}$ is of order of $\dfrac{-1}{S(0)} \sim -13$. This value may not be unique because of the thermodynamic inconsistency discussed in Chapter 4, section 4.3. The relation (A5.7) again holds however in the Born–Green theory.

Finally, we stress that this argument is not applicable to liquid metals. Whereas for liquid insulators like argon, the theory discussed above shows that $S(0)$ plays a dominant role, $S(2k_f)$ will be expected to replace it in metals (cf. section 5.4). Since this is $\sim$ unity, unlike $S(0)$, we do not expect the Born–Green theory to differ so violently from the other theories in its asymptotic predictions for liquid metals.

Conductivity and van Hove correlation function

BAYM (1964) has drawn attention to the fact that the inelastic scattering of a slow neutron from a metal is a process which is very similar to the inelastic scattering of a conduction electron from the lattice vibrations in a metal. In each case, the coupling is not to individual phonons, but to the ion density and Baym's claim is that in each case the Born approximation may be used. This, in the present writer's view, is not established with any certainty for conduction electron scattering, but we shall nevertheless assume it in what follows. While Baym's argument will lead us to the basic formula (7.40), it is most readily stated in terms of the crystal rather than the liquid at the outset.

Then the Bloch electrons, in a metal with rigid ion cores, are scattered via a screened potential by the fluctuations of ionic density. The scattering probability is thus proportional to the number $S'(K, \omega)$ of available states for the density fluctuations. This number is just the van Hove correlation function $S(K, \omega)$ with the Bragg peaks subtracted out.

Now we turn to the usual Boltzmann equation formulation (cf. Chapter 7, sections 7.2 and 7.3). In a collision with the lattice, an electron in a Bloch state $\mathbf{k}$ with energy $E_{\mathbf{k}}$ is scattered into a state $\mathbf{p}$ with energy $E_{\mathbf{p}}$ by creating a density fluctuation with momentum $\mathbf{K}$ and energy ω, putting $\hbar = 1$. Suppose the matrix element of the screened potential is $\langle \mathbf{p} \, U(K, \omega) \, \mathbf{k} \rangle$ for this pro-

Appendix 6

cess. Then the rate at which the scattering occurs is given as usual by

$$2\pi\delta(E_p - E_k - \omega) f(\mathbf{k}) [1 - f(\mathbf{p})] \varrho_0 S'(K, \omega) |\langle \mathbf{p}U(K, \omega)\mathbf{k}\rangle|^2 \quad (A6.1)$$

where $f(\mathbf{k})$ is the density of electrons in the Bloch state $\mathbf{k}$, $1 - f(\mathbf{p})$ is the density of available final states $\mathbf{p}$ and ϱ_0 is the ionic density. The collision term may then be written

$$\left(\frac{\partial f(\mathbf{k})}{\partial t}\right)_{\text{collision}} = \sum_{\mathbf{p}K} \int_{-\infty}^{\infty} d\omega\delta(E_p - E_k - \omega) \varrho_0 |\langle \mathbf{p}U(K, \omega)\,\mathbf{k}\rangle|^2$$
$$\times [f(\mathbf{p})\{1 - f(\mathbf{k})\} S'(-K, -\omega) - f(\mathbf{k})\{1 - f(\mathbf{p})\} S'(K, \omega)].$$
$$(A6.2)$$

The identification of S (minus the elastic Bragg peaks) with the experimentally observed equilibrium density of states for lattice fluctuations assumes that the phonon rate of approach to equilibrium is much more rapid than electron–phonon scattering rates. We must add the condition of detailed balance

$$S'(-K, -\omega) = e^{-\beta\omega}S'(K, \omega) \; ; \beta = (k_B T)^{-1}. \quad (A6.3)$$

If we now use the customary variational calculations with the Boltzmann equation, in which we linearize the collision term, assume free electrons with effective mass m, and take the matrix element to depend only on $\mathbf{K}$;

$$|\langle \mathbf{p} | U(K, \omega) | \mathbf{k}\rangle|^2 \rightarrow |U(K)|^2, \quad (A6.4)$$

then the conductivity is given by

$$\sigma = \frac{n_0 e^2}{m}\tau, \quad (A6.5)$$

where

$$\frac{1}{\tau} = \frac{m}{12\pi^3 Z}\int_0^{2k} dKK^3 |U(K)|^2 \int_{-\infty}^{\infty} \frac{d\omega}{2\pi}\frac{S'(K, \omega)\beta\omega}{[e^{\beta\omega} - 1]}$$
$$(A6.6)$$

where Z is the valence.

This equation was also derived by Mannari (1961) for the special case of a liquid metal. This formula goes into the basic resistivity result of eqn. (7.40) since $\beta\omega/(e^{\beta\omega} - 1)$ can then be replaced by unity and the integral over ω obviously gives us the structure factor $S(K)$. Baym points out that, in consequence, the Ziman formula includes all multiphonon processes.

Notes added in proof

1. A. S. Marwaha and N. E. Cusack (1966: *Phys. Letters* **22**, 556) have recently reported more detailed measurements of the absolute thermopower Q. In general, their results are in reasonable agreement with those given in Table 3, p. 71. The missing entry for Tl is thereby completed, and agreement with Sundström's calculations is fair.

2. Using the theory given in Chapter 7, § 7.11, and X-ray partial structure factors for liquid Ag-Sn alloys, N. C. Halder and C. N. J. Wagner (1967: *Phys. Letters* **24A**, 345–6) have shown that the electrical resistivity is in good agreement with experiment over the entire range of concentration.

3. In a very recent paper, the work of Watabe and Tanaka on Zn, and in particular the density of states curve shown in Fig. 30, p. 106 has been severely criticized by L. E. Ballentine (1966: *Canadian Journal of Physics* **44**, 2533–52). Ballentine's work is still based on the theory of Edwards which is given in Chapter 9, but his work differs from that of Watabe and Tanaka in two respects:

(i) Their screened Coulomb pseudo-potential (9.22) is replaced by the model potential of Heine and Abarenkov, discussed in Chapter 7, § 7.6.

(ii) Instead of using simply the real part of $\sum(\mathbf{k}, E)$, Ballentine calculates the imaginary part also, and hence the density of states $\langle n(E) \rangle$ from (9.3), whereas Watabe and Tanaka replace $\varrho(E, \mathbf{k})$ by a delta function, and use the approximation (9.20) to calculate $\langle n(E) \rangle$.

However, according to Ballentine, their procedure should still lead to a much more free-electron-like density of states curve for Zn than that shown in Fig. 30. The treatment of the singularities in the logarithm in eq. (9.19) needs especial care.

The main conclusions of Ballentine are that the electronic structures of Al and Zn are nearly free-electron-like. For Bi, the other liquid metal he considered in detail, the density of states appears to be much more complicated, and the Edwards theory is probably inappropriate.

References

ALDER, B. J. (1964) *Phys. Rev. Letters* **12**, 317–19.
ALFRED, L. C. R., and MARCH, N. H. (1957) *Phil. Mag.* **2**, 985–97.
ASCARELLI, P. (1966) *Phys. Rev.* **143**, 36–47.
ASHCROFT, N. W. and LEKNER, J. (1966) *Phys. Rev.* **145**, 83–90.
ASHCROFT, N. W. and MARCH, N. H. (1967) *Proc. Roy. Soc.* **A297**, 336–50.
BARDEEN, J. (1937) *Phys. Rev.* **52**, 688–97.
BARDEEN, J. and PINES, D. (1955) *Phys. Rev.* **99**, 1140–50.
BAYM, G. (1964) *Phys. Rev.* **135**, A1691–2.
BLANDIN, A., DANIEL, E. and FRIEDEL, J. (1959) *Phil. Mag.* **4**, 180–2.
BOHM, D. and STAVER, T. (1952) *Phys. Rev.* **84**, 836–7.
BORN, M. and GREEN, H. S. (1946) *Proc. Roy. Soc.* **A188**, 10–18.
BRAGG, W. L. and WILLIAMS, E. J. (1934) *Proc. Roy. Soc.* **A145**, 699–730.
CATTERALL, J. A. and TROTTER, J. (1963) *Phil. Mag.* **8**, 897–902.
CHEN, S. H., EDER, O. J., EGELSTAFF, P. A., HAYWOOD, B. C. G. and WEBB,
    F. J. (1965) *Phys. Letters* **19**, 269–71.
COCHRAN, W. (1965) *Inelastic Scattering of Neutrons*, 1 (Vienna: Int. Atomic
    Energy Agency), p. 3.
COMPTON, A. H. and ALLISON, S. K. (1935) *X-rays in Theory and Experiment*,
    Van Nostrand, Princeton.
CORLESS, G. K. and MARCH, N. H. (1961) *Phil. Mag.* **6**, 1285–96.
COWLEY, R. A., WOODS, A. D. B. and DOLLING, G. (1966) *Phys. Rev.* **150**,
    487–94.
CUSACK, N. E. (1963) *Reports on Prog. Phys.* **26**, 361–409.
DANIEL, E. and VOSKO, S. H. (1960) *Phys. Rev.* **120**, 2041–44.
DEBYE, P. (1915) *Ann. d. Physik* **46**, 809–23.
DE GENNES, P. G. (1959) *Physica* **25**, 825–39.
DESAI, R. C. and NELKIN, M. (1966) *Phys. Rev. Letters* **16**, 839–41.
EDWARDS, S. F. (1962) *Proc. Roy. Soc* **A267**, 518–40.
EGELSTAFF, P. A., DUFFILL, C., RAINEY, V., ENDERBY, J. E. and NORTH,
    D. M. (1966) *Phys. Letters* **21**, 286–8.
ENDERBY, J. E. (1963) *Proc. Phys. Soc.* **81**, 772–9.
ENDERBY, J. E. and MARCH, N. H. (1965) *Adv. Phys. (Phil. Mag. Suppl.)* **14**,
    453–77.
    (1966a) Battelle Conference on Phase Stability, Geneva. To appear in
    Proceedings.
    (1966b) *Proc. Phys. Soc.* **88**, 717–21.
ENDERBY, J. E., GASKELL, T. and MARCH, N. H. (1965) *Proc. Phys. Soc.* **85**,
    217–21.

126

References

ENDERBY, J. E., NORTH, D. M. and EGELSTAFF, P. A. (1966) *Phil. Mag.* **14**, 961–70.

ENDERBY, J. E., TITMAN, J. M. and WIGNALL, G. D. (1964) *Phil. Mag.* **10**, 633–40.

FABER, T. E. (1966) *Optical Properties and Electronic Structure of Metals and Alloys*, Ed. F. Abelès, North Holland, Amsterdam.

FABER, T. E. and ZIMAN, J. M. (1965) *Phil. Mag.* **11**, 153–73.

FEYNMAN, R. P. and COHEN, M. (1956) *Phys. Rev.* **102**, 1189–204.

FRANK, F. C. (1939) *Proc. Roy. Soc.* **A170**, 182–9.

FRISCH, H. L. and LEBOWITZ, J. L. (1964) *The Equilibrium Theory of Classical Fluids*, Benjamin, New York.

FUMI, F. G. (1955) *Phil. Mag.* **46**, 1007–20.

GAMERTSFELDER, C. (1941) *J. Chem. Phys.* **9**, 450–7.

GASKELL, T. (1965) *Proc. Phys. Soc.* **86**, 693–5.

(1966) *Proc. Phys. Soc.* **89**, 231–6.

GASKELL, T. and MARCH, N. H. (1963) *Phys. Letters* **7**, 169–70.

GINGRICH, N. S. (1943) *Rev. Mod. Phys.* **15**, 90–110.

GINGRICH, N. S. and HEATON, L. (1961) *J. Chem. Phys.* **34**, 873–8.

GLAUBER, R. J. (1955) *Phys. Rev.* **98**, 1692–8.

GREEN, H. S. (1952) *Molecular Theory of Fluids*, North Holland, Amsterdam.

GREENE, M. P. and KOHN, W. (1965) *Phys. Rev.* **137**, A513–22.

GREENFIELD, A. J. (1966) *Phys. Rev. Letters* **16**, 6–8.

GUBANOV, A. I. (1965) *Quantum Electron Theory of Amorphous Conductors*, Consultants Bureau, New York.

GUSTAFSON, D. R., MACKINTOSH, A. R. and ZAFFARANO, D. J. (1963) *Phys. Rev.* **130**, 1455–9.

HARRISON, W. A. (1963) *Phys. Rev.* **129**, 2512–24, ibid. **131**, 2433–42.

(1964) *Phys. Rev.* **136**, A1107–19.

(1965) *Phys. Rev.* **139**, A179–85.

(1966) *Pseudopotentials in the Theory of Metals*, Benjamin, New York.

HEINE, V. and ABARENKOV, I. (1964) *Phil. Mag.* **9**, 451–65.

HENSHAW, D. G. (1957) *Phys. Rev.* **105**, 976–81.

(1958) *Phys. Rev.* **111**, 1470–5.

JOHNSON, M. D. and MARCH, N. H. (1963) *Phys. Letters* **3**, 313–14.

JOHNSON, M. D., HUTCHINSON, P. and MARCH, N. H. (1964) *Proc. Roy. Soc.* **A282**, 283–302.

KIRKWOOD, J. G. (1935) *J. Chem. Phys.* **3**, 300–13.

KITTEL, C. (1963) *Quantum Theory of Solids*, Wiley, New York.

KOHN, W. and VOSKO, S. H. (1960) *Phys. Rev.* **119**, 912–18.

KRISHNAN, K. S. and BHATIA, A. B. (1945) *Nature, London* **156**, 503–4.

KUBO, R. (1964) *Physics of Semiconductors* (Proceedings of 7th International Conference), Dunod, Paris.

LANDAU, L. D. and LIFSHITZ, E. M. (1958) *Statistical Physics*, Pergamon, Oxford.

LANGER, J. S. and VOSKO, S. H. (1959) *Jour. Phys. Chem. Solids* **12**, 196–205.

LAZARUS, D. (1954) *Phys. Rev.* **93**, 973–6.

LENNARD-JONES, J. E. and DEVONSHIRE, A. F. (1939) *Proc. Roy. Soc.* **A170**, 464–84.

127

References

LIGHTHILL, M. J. (1958) *An Introduction to Fourier Analysis and Generalised Functions*, Cambridge University Press, London.

LINDE, J. O. (1932) *Ann. Phys. Lpz.* **15**, 219–48.

LINDHARD, J. (1954) *Kgl. Danske Mat.-fys. Medd.*, **28**, 8.

MAKINSON, R. E. B. and ROBERTS, A. P. (1960) *Aust. J. Phys.* **13**, 437–45.

MANNARI, I. (1961) *Prog. Theor. Phys.* **26**, 51–83.

MARCH, N. H. (1966) *Phys. Letters* **20**, 231–2.

MARCH, N. H. and MURRAY, A. M. (1960) *Phys. Rev.* **120**, 830–6.
(1961) *Proc. Roy. Soc.* **A261**, 119–33.
(1962) *Proc. Roy. Soc.* **A266**, 559–67.

MARCH, N. H., YOUNG, W. H. and SAMPANTHAR, S. (1967) *The Many-Body Problem in Quantum Mechanics*, Cambridge University Press.

MESSIAH, A. (1961) *Quantum Mechanics*, North Holland, Amsterdam.

MEYER, A., NESTOR, C. W. and YOUNG, W. H. (1965) *Phys. Letters* **18**, 10–11.

MIKOLAJ, P. G. and PINGS, C. J. (1967) *J. Chem. Phys.* **46**, 1412–20.

MOTT, N. F. (1936) *Proc. Camb. Phil. Soc.* **32**, 281–90.

MOTT, N. F. (1966) *Phil. Mag.* **13**, 989–1014.

MOTT, N. F. and JONES, H. (1936) *Theory of the Properties of Metals and Alloys*, Clarendon Press, Oxford.

MUKHERJEE, K. (1965) *Phil. Mag.* **12**, 915–18.

NELKIN, M. S. and PARKS, D. E. (1960) *Phys. Rev.* **119**, 1060–8.

NIJBOER, B. A. and RAHMAN, A. (1966) *Physica*, **32**, 415–32.

NIJBOER, B. R. A. and VAN HOVE, L. (1952) *Phys. Rev.* **85**, 777–83.

NORDHEIM, L. (1931) *Ann. Phys. Lpz.* **9**, 641–78.

ORTON, B. R., SHAW, B. A. and WILLIAMS, G. I. (1960) *Acta Metall.* **8**, 177–86.

PAALMAN, H. H. and PINGS, C. J. (1963) *Rev. Mod. Phys.* **35**, 389–99.

PASKIN, A. and RAHMAN, A. (1966) *Phys. Rev. Letters* **16**, 300–3.

PERCUS, J. K. and YEVICK, G. J. (1958) *Phys. Rev.* **110**, 1–13.

RAIMES, S. (1961) *Wave Mechanics of Electrons in Metals*, North Holland, Amsterdam.

RAHMAN, A. (1964) *Phys. Rev.* **136**, A405–11.

RANDOLPH, P. R. (1964) *Phys. Rev.* **134**, A1238–48.

ROWLINSON, J. S. (1965) *Reports on Progress in Physics* **28**, 169–99.

RUSHBROOKE, G. S. (1960) *Physica* **26**, 259–65.

SCHOFIELD, P. (1961) *Proc. Int. Atomic Energy Agency Symposium: Vienna*, p. 39.

SEYMOUR, E. F. H. and STYLES, G. A. (1964) *Phys. Letters* **10**, 269–70.

SIMMONS, R. O. and BALUFFI, R. W. (1960) *Phys. Rev.* **117**, 52–61.

SPRINGER, B. (1964) *Phys. Rev.* **136**, A115–24.

STEWART, A. T., KUSMISS, J. H. and MARCH, R. H. (1963) *Phys. Rev.* **132**, 495–7.

STOTT, M. J. and MARCH, N. H. (1966) *Phys. Letters* **23**, 408–9.

SUNDSTRÖM, L. J. (1965) *Phil. Mag.* **11**, 657–65.

THIELE, E. (1963) *Jour. Chem. Phys.* **39**, 474–9.

VAN HOVE, L. (1954) *Phys. Rev.* **95**, 249–62.

VINEYARD, G. H. (1958) *Phys. Rev.* **110**, 999–1010.

VOSKO, S. H., TAYLOR, R. and KEECH, G. H. (1965) *Can. Jour. Phys.* **43**, 1187–247.

References

WARREN, B. E. and GINGRICH, N. S. (1934) *Phys. Rev.* **46**, 368–72.
WATABE, M. and TANAKA, M. (1964) *Prog. Theor. Phys.* **31**, 525–37.
WERTHEIM, M. S. (1963) *Phys. Rev. Letters* **10**, 321–3.
WISER, N. (1966) *Phys. Rev.* **143**, 393–8.
WORSTER, J. and MARCH, N. H. (1964) *Solid State Communications*, **2**, 245–7.
ZERNIKE, F. and PRINS, J. A. (1927) *Z. Phys.* **41**, 184–94.
ZIMAN, J. M. (1960) *Electrons and Phonons*, Oxford University Press.
(1961) *Phil. Mag.* **6**, 1013–34.
(1964) *Adv. Phys. (Phil. Mag. Suppl.)* **13**, 89–138.

129

Index

Index

PRINTED IN HUNGARY
Franklin Printing House, Budapest